Exploring a New Congestion Pricing Concept:
Focus Group Findings from Northern Virginia and Philadelphia

Prepared by:

Margaret Petrella, Lee Biernbaum, and Jane Lappin
Volpe National Transportation Systems Center
Cambridge, MA

Prepared for:

Federal Highway Administration
Office of Natural and Human Environment and Office of Transportation Management

December 2007

Table of Contents

Executive Summary i
Background and Study Objectives 1
Study Design 3
Findings 8
Conclusions and Recommendations 28
Appendix A: Discussion Guides 32
Appendix B: Scenarios 41
Appendix C: Worksheets on Transportation Costs and Toll Costs 45
Appendix D: Questions Raised by Participants 53

Executive Summary

In support of the U.S Department of Transportation's (DOT) National Strategy to Reduce Congestion on America's Transportation Network, the Federal Highway Administration's (FHWA) Office of Natural and Human Environment (HEPN) and Office of Transportation Management (HOTM) sought assistance from The Volpe National Transportation Systems Center to explore attitudes about congestion pricing. For this exploratory study, focus groups were convened in Northern Virginia and Philadelphia among the general public, business owners and managers, and owners and managers of shipping and transportation logistics firms. The purpose of these focus groups was to obtain feedback on a specific congestion pricing scenario and to better understand the public's concerns regarding congestion pricing. A secondary purpose was to learn more about how to communicate with the public on the topic of congestion pricing.

Each of the focus groups opened with a discussion of the transportation issues or concerns that participants face in their everyday life. A variety of issues were raised but traffic congestion dominated the discussion. Respondents were then asked if they knew how transportation is funded. Many were aware that the gas tax funds transportation, but few could name other sources of transportation revenue. When asked to comment on the current tax-based system for funding transportation, respondents did not have strong feelings on the topic, instead they tended to express dissatisfaction with the way government spends their tax dollars and the perceived lack of results.

At this point in the discussion, participants were asked to read a handout, or scenario, that described the user-based congestion pricing system being evaluated for this study. Respondents were asked to consider the possibility that their current tax-based system for funding transportation would be replaced by a user-based congestion pricing system. That is, the portion of their taxes that is collected to fund transportation would be eliminated, and instead transportation would be funded through charging tolls on highways during peak hours of travel, with tolls set high enough to allow for free flow traffic. In both Northern Virginia and Philadelphia the scenario indicated that the toll would range from 25 cents during peak hours to 0 cents during off peak, and the tolls would be implemented in major metropolitan areas across the country.

As described in the scenario, tolls would be collected using electronic toll technology (i.e., E-ZPass), and drivers without transponders would be video-tolled. In addition, tolls would be adjusted periodically based on changes in traffic patterns in order to ensure free-flow traffic. In combination with tolling, employers would be encouraged to provide their employees with telecommuting and flextime options, public transportation would be expanded to include a new express bus service, and new park and ride lots would be built to encourage carpools and vanpools (certified vanpools would not pay the toll and carpools could split the toll).

Overall there was a mix of opinion regarding the proposed congestion pricing system. Within each group, some respondents thought it might work, even though they had concerns about certain aspects of the system. (One such respondent referred to himself as a "marginal supporter.") Others were opposed to the system, indicating for example, that it would not work or that it was too costly. Most respondents, however, were lukewarm or unsure about the concept; they were willing to consider it, despite having reservations about it. As one such respondent indicated, "I'm stuck on neutral on this."

The shipper groups appeared somewhat more open to the idea of congestion pricing than the business owners were, perhaps as some noted, because shippers can pass the costs on to their customers. Also, some respondents in the shipper groups said they would schedule their travel during off-peak hours (though courier services and taxi and limousine services said they do not have this flexibility). The business groups appeared to be somewhat more sensitive to the costs of the new system. For many businesses, the location of their office and/or the nature of their business meant that employees were constantly on the highway, either commuting to or from work or meeting with clients during the day. Consequently, they felt the congestion pricing concept would have an adverse economic impact. A number of business respondents also indicated that they couldn't offer telecommuting or flextime to their employees due to the nature of their business.

In discussing the congestion pricing scenario, many of the same concerns and worries were raised in both cities and across the three types of groups. The concerns raised most frequently included:

- *Diversion to arterials:* In all the groups, a key concern was that local roads would become extremely more congested.

- *Administrative costs:* Respondents perceived the system would require high start-up costs to outfit the highways and to equip drivers with transponders. Respondents also felt that video-tolling, specifically tracking down drivers who do not have a transponder, would result in significant expense. This was of particular concern regarding visitors or tourists; respondents wanted to be sure that visitors paid for their use of the roads (and in fact some thought visitors should be charged more than residents).

- *Distrust of government:* Respondents in both Northern Virginia and Philadelphia seemed to distrust government to effectively administer the new program. Many respondents voiced doubts that the existing taxes that fund transportation would really be eliminated. Others questioned whether or not the government would use the funds collected from tolls for transportation improvements, and respondents (particularly in Philadelphia) also worried that the government would simply keep raising the price of the toll.

- *High personal (business) cost:* Respondents in both cities and across the different types of groups worried that the system could be a financial burden and that it would cost them more than they are currently paying in taxes.

- *Equity issues:* Equity issues took on several different guises during the group discussions. Some (more so in Northern Virginia than Philadelphia) were concerned about the ability of low income people to pay the tolls. Respondents also mentioned that it was unfair that commuters (especially those with less flexible work schedules) would be paying to fund the roads, whereas off-peak drivers would not have to pay anything.

- *Enforcement difficulties:* Respondents questioned government's ability to enforce the system and predicted that people would try to "cheat" the system. One concern with low-income discounts, for example, was that people would begin driving their grandmothers' car.

- *Won't solve the problem:* A number of respondents simply could not fathom that highways would be free of congestion, despite admission by some that they would consider changing their schedule, mode, or route if the system were implemented. Many respondents did not seem to make the connection that these changes in behavior would result in less congestion on the highways.

- *Restraint of choice:* Respondents want the choice of whether or not they pay for their use of the roads. With the congestion pricing scenario, some felt like they were being pushed off the roads. Others indicated that if public transportation were improved, then they would have a choice; otherwise congestion pricing was perceived as a "forced" cost.

- *Privacy:* Privacy was raised in nearly all of the focus groups, however, it was not an issue that resonated strongly or generated much discussion, with the exception of one or two respondents who seemed to care deeply about this issue.

Other concerns were raised by fewer respondents. These included the following:

- Toll revenues will be insufficient to fund all roads
- System is unfair to highway drivers
- System is punishment
- New rush hour will be created as drivers seek to avoid the toll charged during peak hours
- Seniors will be opposed
- Geographic equity issues will arise (if the system is implemented in metropolitan areas only)
- Businesses will have difficulty budgeting for the tolls
- Jurisdictional disputes will arise (i.e., Federal vs. State authority)

Despite their concerns, respondents were also able to articulate the potential benefits of the new concept, including reduced congestion, time savings and reduced emissions. In particular, respondents reacted very positively toward E-ZPass, and the general public was generally favorable towards telecommuting and flextime, with some respondents indicating that they would take advantage of such options (though others pointed out that telecommuting or flextime were not feasible in their business). While some mentioned that the new system would provide an incentive to carpool, most did not seem willing to carpool themselves. Some respondents, however, did indicate they would switch to public transportation if service were expanded to their area. In Philadelphia respondents felt that for transit to become a viable option, there would need to be significant improvements in reliability, safety and cleanliness.

Overall, respondents wanted more information about the mechanics of the new system and evidence that it would in fact reduce congestion. Many were hopeful that congestion pricing might help solve the problem of traffic congestion, and a number thought that the new system was worth a try. People's willingness to consider a congestion pricing system seems to be due, in part, to their sense that traffic congestion has only gotten worse over time and that "something has to be done."

Background and Study Objectives

In support of the U.S Department of Transportation's (DOT) National Strategy to Reduce Congestion on America's Transportation Network, the Federal Highway Administration's (FHWA) Office of Natural and Human Environment (HEPN) and Office of Transportation Management (HOTM) sought assistance from The Volpe National Transportation Systems Center to explore attitudes about congestion pricing. For this exploratory study, focus groups were convened in Northern Virginia and Philadelphia. The purpose of these focus groups was to obtain feedback on a specific congestion pricing scenario and to better understand the public's questions and concerns about congestion pricing. A secondary purpose was to learn more about how to communicate with the public on the topic of congestion pricing.

In previous research, focus groups have explored congestion pricing more generally, with many focusing on High Occupancy Tolling. Overall, these studies find that the public is split on tolling (some see it as fair because users pay; others feel it is unfair due to "double taxation"). Value pricing is generally more acceptable on new facilities than existing ones, and the availability of a "free option" has arisen as a key factor in user acceptability. Earlier studies reveal that the public has concerns about equity, as well as questions about the technology and enforcement. Interestingly, a number of studies have found that initial skepticism about value pricing tends to be overcome once an area adopts tolling.

For this project, a new congestion pricing concept was being evaluated as a means of easing traffic congestion on the nation's highways. Respondents were asked to consider the possibility that their current tax-based system for funding transportation would be replaced by a user-based congestion pricing system. That is, the portion of their taxes that is collected to fund transportation would be eliminated, and instead transportation would be funded through charging tolls on highways during peak hours of travel, with tolls set high enough to allow for free flow traffic. In both Northern Virginia and Philadelphia the scenario indicated that the toll would range from 25 cents during peak hours to 0 cents during off peak, and the tolls would be implemented in major metropolitan areas across the country.

The scenario also described the following components of the new system:

- The tolls would be collected using electronic toll technology (i.e., E-ZPass), and drivers without transponders would be video-tolled
- Tolls would be adjusted periodically based on changes in traffic patterns in order to ensure free-flow traffic
- Employers would be encouraged to provide their employees with telecommuting and flextime options
- Public transportation would be expanded to include a new express bus service

- New park and ride lots would be built to encourage carpools and vanpools (certified vanpools would not pay the toll and carpools could split the toll)

As part of this effort, FHWA wanted to explore the attitudes and opinions of a range of stakeholders. In addition to convening focus groups among the general public, separate groups were also conducted among: 1. business owners and/or managers (referred to as "business groups", and 2. owners and/or managers of shipping and transportation logistics firms (referred to as "shipper groups"). It was hypothesized that the business and shipper groups might have different issues and concerns from the general public, so it would be useful to conduct separate groups to obtain their perspectives.

It is worth noting that the purpose of focus groups, a form of qualitative research, is to develop insight, in-depth understanding of attitudes and behavior, and direction for planning or further research. Because of the small number of people interviewed and the non-random method of recruitment, the findings cannot be analyzed quantitatively, nor are they necessarily representative of the general population.

Study Design

<u>Site Selection</u>

For the focus groups, Northern Virginia and Philadelphia were identified as sites for the study. Both are large urban centers on the east coast with well-subscribed transit service. Despite these similarities, the differences between the sites made them interesting candidates for the study. Northern Virginia has experienced significant growth over the last decade, whereas Philadelphia has seen relatively low growth and has an aging infrastructure. In addition, based on data from the Texas Transportation Institute, the Washington D.C. metro area has greater traffic congestion and delay relative to Philadelphia.

Of the four focus groups convened at each site, two consisted of the general public, one was held among small business owners and managers, and another group was comprised of owners and managers of shipping and/or transportation logistics firms.

The focus groups in Northern Virginia were convened on July 11 and 12, 2007 at a focus group facility in Fairfax, Virginia. In Philadelphia, the focus groups were convened on July 23 and 24, 2007. The general public groups were held at a focus group facility in Bala Cynwyd, Pennsylvania (a suburb of Philadelphia), while the business and shipper groups were conducted in center city Philadelphia.

<u>Facilitation</u>

A professional moderator, Linda LaScola, of LaScola Qualitative Research, led the discussion for all eight focus groups. With postgraduate training in psychotherapy and group dynamics and over 20 years of experience in conducting focus groups, the moderator brought significant value to the project. Her training and experience enabled a balanced and unbiased exploration of the issues.

The moderator was involved in initial planning meetings for the focus groups and provided input on the design of both the recruitment screeners and the discussion guides. In addition, the moderator wrote summary comments on findings from the groups and provided feedback on the draft report.

<u>Focus Group Screeners</u>

Separate screener questionnaires were designed to recruit participants for each of the focus group types, including the general public, business owners and the shipper/transportation logistics group. The following section highlights the screening criteria for each group.

General Public

In order to obtain a mix of respondents with different demographic and socioeconomic characteristics, quotas were used. A balanced distribution was sought on the following measures, in order to ensure that people with different travel characteristics and needs were being represented:

- Gender
- Age
- Education
- Income
- Household composition (with and without children)
- Employment status (retired, employed, homemakers)

The screener included several eligibility requirements. For the four public groups, respondents had to meet the following requirements:

- Must have driver's license
- Must drive vehicle at least 3-5 days/week
- Must drive between 6,000 -16,000 miles per year
- Must drive on the highway

While all the respondents had to be drivers, effort was made to obtain variation on the following measures (it was hypothesized that these variables would be related to attitudes about congestion pricing, so recruiting a mix of respondents would insure that a range of opinions was represented):

- Peak hour driving (at least 5 drive on the highway during peak hours 3-5 days/week or more; 2-3 drive on the highway during peak hours less than one day a week)
- Toll payments (2 -3 pay tolls daily; at least 3 pay tolls "less than once a month" or "never.")
- E-ZPass (Smart Tag) ownership (at least 2 and no more than 5 have an E-ZPass)

The screener also included criteria regarding the participants' professional background. Eligible participants could not work for a market research firm, the media (including print, TV, radio, or public relations or advertising in the media), public transportation agency or authority, Department of public works; local, state or Federal Department of Transportation; taxi or commercial driver; an elected or appointed official; or a public safety official. Such participants could be perceived as "experts," leading to an unproductive group dynamic. Previous focus group participation was also taken into consideration, to eliminate "professional" focus group participants. Those who had participated in a focus group within the last six months were ineligible.

Business and Shipping/Transportation Logistics

For the business and shipper groups, owners or managers were asked to participate. The aim was to obtain the perspectives of persons in the business most responsible for making executive decisions, as they would be best informed about the business and its finances, and could speak to the impact of congestion pricing on both their business and their employees.

For the business group, the screener included questions designed to recruit a mix of business types, such as consulting companies, hotels, restaurants, retail stores, and supermarkets. In addition, variation was sought on the size of business, so that both smaller and larger businesses would be included in the group. The following table presents the different business types that participated in each city.

Northern Virginia	Philadelphia
<ul><li>IT consulting</li><li>retail store (art framing)</li><li>restaurant/catering</li><li>Tourism and Convention Bureau</li><li>strategic consulting</li><li>restaurant/ice cream shop</li><li>Chamber of Commerce</li></ul>	<ul><li>Hotel</li><li>business/IT consulting</li><li>construction</li><li>detective agency</li><li>wholesale baker</li><li>restaurant/catering</li><li>catering</li><li>architecture</li></ul>

Likewise, for the shipper groups, the screener was designed to recruit a mix of types, including:

- Utilities (phone, cable, gas, electric)
- Delivery services (Federal Express/UPS/local couriers)
- Local service or trucking fleets
- Regional trucking firms
- Long distance trucking
- Local for hire operators (taxi firms, limousines)

The focus group screener also sought to obtain a mix of fleet size, in order to represent the viewpoints of both smaller and larger companies. The types of shipping/transportation logistics firms that participated in the study are outlined in the table below (see next page).

Northern Virginia	Philadelphia
<ul><li>petroleum hauling</li><li>overnight delivery services</li><li>truck leasing/rental</li><li>limousine operator</li><li>heat and air conditioning repair service</li><li>limousine service</li><li>courier/transportation logistics</li></ul>	<ul><li>moving company</li><li>moving company</li><li>container mover</li><li>truck leasing/rental</li><li>limousine operator</li><li>taxi/limousine operator</li><li>delivery service (same day)</li></ul>

Focus Group Discussion Guides[1]

The discussion guide questions were designed to elicit the public's candid perspectives about the transportation issues that matter to them, about the current tax based system for funding transportation, and about a new user-based congestion pricing system. The conversation began by exploring the group's more general views about the transportation issues that matter to them. As anticipated, traffic congestion was a topic raised in all the groups, and respondents were prompted to discuss whether this problem was solvable. In order to gauge respondent awareness of the current tax-based system for funding transportation, respondents were asked whether or not they knew how transportation is funded (i.e., the specific funding sources). Following this discussion, the moderator presented them with a list of the different sources, and invited them to share their thoughts about the current system for funding transportation.

After that participants were asked to read a handout, or scenario, that described the user-based congestion pricing system being evaluated for this study. Based on feedback from the first set of focus groups in Northern Virginia that the scenario was overly favorable towards the new system, the scenario was rewritten to be more balanced and was formatted as a newspaper article for the Philadelphia focus groups (see Appendix B for the scenarios).

In addition to reading the scenario, respondents were given a map of the area that highlighted the highways that would be tolled under the new congestion pricing system. The map provided a common frame of reference for all participants, and was a useful tool for ensuring that they understood which roads were being tolled.

After reading the scenario, respondents were asked about their impressions of the new concept. As part of this discussion, some of the following topics were covered:

- What concerns you about this user-based congestion pricing system?
- What are the potential benefits?
- Under what conditions would you accept a toll based system?

---

[1] See Appendix A for the focus group discussion guides.

In order to make the system more tangible or realistic, worksheets were developed that enabled participants to calculate what their tax savings would be (i.e., if they no longer had to pay the taxes that fund transportation). Similarly, they were given a worksheet that led them through the steps necessary to calculate what they would pay in tolls under the new system, based on their current driving habits (See Appendix C for Worksheets). Comparing their tax savings to their toll costs gave respondents a better sense of how the new system would affect them personally. Again, reactions to the new system were elicited, based on this more personal understanding.

The discussion guide used for the business and shipper/transportation logistics firms was very similar to the one used by the general public, except that the business and shipper groups were prompted to think about these issues from the perspective of their business. Another notable difference is that the business and shipping groups were not given worksheets to calculate their tax savings and potential toll costs. Developing such worksheets for businesses would have been extremely complicated, as taxes vary across industries. It was anticipated that business and shipping owners/managers would have a good sense of their current transportation costs, and could more easily estimate the impact of the new congestion pricing system on their business (i.e. without having to complete the worksheets).

Findings

This section highlights key findings from all eight focus groups. Where relevant, distinctions are drawn between the general public and the business and shipper groups. The following is an outline of the key findings:

- **Transportation issues: "Traffic dictates how we live our lives"**
- **How is transportation funded: general awareness, but short on specifics**
- **Groups don't have strong opinions on the current tax-based system for funding transportation; but they have lots to say about the *way* funds are spent**
- **Congestion-based pricing: they get it, for the most part**
- **What's the purpose?**
- **Congestion-based pricing: a mixed reaction**
- **People want more information on the program; in particular they want to know what their tax savings would be and how much they would be paying in tolls**
- **Congestion-based pricing system raises a variety of concerns**
- **In an "alternate universe," the public perceives the benefits of congestion pricing**
- **Some would change their travel behavior, or where they live and work**
- **Telecommuting, flextime, carpools, and express bus service have some appeal**
- **E-ZPass is popular**
- **Environmental benefits not on most people's radar, though potential benefits acknowledged**
- **Consider these improvements…**
- **There are other ways to fix the problem**

Transportation issues: "Traffic dictates how we live our lives." (public, Northern Virginia)

In both Northern Virginia and Philadelphia, the discussion opened with a general question about what transportation issues affect participants [and/or their business] in their everyday life. Traffic congestion was usually the first issue raised in each of the groups, and it was the issue that generated the most discussion. Respondents voiced concerns about the amount of delay they experience due to traffic and the lack of predictability.

- "Unpredictability at all times of day…[you] don't know how long it will take to get someplace." (public, Northern Virginia)
- "Fairfax County Parkway is a parking lot." (business, Northern Virginia)
- "Too many people on the road." (public, Philadelphia)

Both the general public groups and the business/shipper groups lamented the impact of traffic congestion on their daily lives and their businesses. One respondent in Northern Virginia pointed out that he doesn't do all the things that he would like to do – such as attend events and festivals – due to the aggravation of dealing with traffic. In Northern Virginia, several of the shipping firms indicated that traffic has affected their business operations; they have had to open additional branches or facilities in order to better serve their customers. In Philadelphia, a construction company indicated that transportation is critical to his business; if the materials don't arrive to the site as scheduled, the job "comes to a stop."

At the same time, respondents were resigned to the fact that congestion is a part of their lives, and they have learned to plan accordingly. Several respondents pointed out that they schedule their day around rush hour traffic and they have learned which routes are the least congested. Business owners explained that they build in extra time between appointments with clients, since "Your client doesn't want hear 'I got stuck in traffic.'"

- "You better schedule what you want to do with plenty of time." (public, Northern Virginia)
- "I plan my days around congestion." (business, Northern Virginia)
- "I plan on-site time to avoid rush hour traffic time." (business, Philadelphia)

This is similar to findings from focus groups conducted in Boise, Seattle and Philadelphia one year ago (July 2006) in which traffic congestion was also a top concern. Respondents in these cities were unhappy about the traffic congestion and the resulting travel delay, but they were somewhat resigned; traffic was a part of their everyday life, and they had developed strategies for dealing with it.

Respondents in both Northern Virginia and Philadelphia were pessimistic when asked if congestion is solvable, indicating that traffic has only been getting worse. In Northern Virginia, respondents noted that people are constantly moving into the region, and so it is difficult to imagine a solution to the problem. In Philadelphia, respondents were more likely to mention the aging infrastructure and "antiquated roads" that were not designed to handle the current volume of traffic. Nonetheless, group participants did offer suggestions for ways to ease traffic congestion. In both cities, there were some respondents who felt that widening or building new roads (adding lanes, building connectors etc.) was the solution, but this suggestion was not supported by all. Others felt that building roads was not a long term solution, as the new roads will simply become congested again, over time.

Other suggestions for easing traffic congestion offered in both cities included telecommuting, flextime, and carpool lanes. However, some respondents, particularly in the business groups, did not view these options as solutions. As one Northern Virginia business owner put it, "we're already doing all these things."

In both cities, respondents agreed that easing traffic congestion required "getting people out of their cars," and they viewed public transportation as a long-term solution.

However, respondents felt that significant transit improvements would be necessary in order for transit to be a feasible solution. In Northern Virginia, (in both the general public and business groups) a couple of respondents cited the need for more extensive service (i.e., "the metro doesn't go where I need it to go"), more parking at metro facilities, and longer hours of operation. In Philadelphia, respondents were significantly more dissatisfied with their public transportation system (compared to Northern Virginia). Participants in the general public, as well as the business and shipper groups, mentioned numerous problems with the service, including reliability, cost, safety, and cleanliness. Residents also complained about recent increases in fares and a new policy requiring that transit riders pay for transfers.

In addition, several other solutions to traffic congestion were mentioned by just one or two respondents, including:

- Raising the price of gas to $5 per gallon (public, Northern Virginia and Philadelphia)
- Charging for HOV lanes (public, Northern Virginia)
- Providing benefits /incentives to carpoolers/transit users (shipper, Northern Virginia)

Other transportation issues

In addition to traffic congestion, the focus groups noted a number of other transportation issues that they face in their everyday life. Gas prices were mentioned by 5 out of the 8 focus groups, including a mix of the general public and business/shipper groups. While some members of the public have altered their travel patterns to conserve gas, others noted that the price of gas has not had much impact on their travel. Businesses mentioned that their costs have increased due to gas prices, and they have had to pass these costs onto their customers.

Other transportation issues raised by relatively fewer respondents included:

- Tolls/Cost of alternative roads (public and business, Northern Virginia; business, Philadelphia)
- Pollution (public, Philadelphia)
- Crazy drivers/road rage (public, Philadelphia)

Several issues were raised specifically in the business or shipping groups:

- Parking (business, Northern VA and Philadelphia)
- Road design (shipper and business groups, Northern Virginia)
- Lack of transportation planning, i.e., "the road system has to be able to support development" (shipper and business, Northern Virginia; shipper, Philadelphia)

- Number of DOT regulations[2] (shipper and business, Philadelphia)

How is transportation funded: general awareness, but short on specifics.

When asked if they knew how transportation was funded, participants only had a general sense of the sources of revenue. Each of the groups mentioned the gas tax, as well as "taxes" more generally, including "state and county" taxes and/or "federal" taxes. Most respondents (particularly the general public, but even to some extent the business and shipper groups), however, do not appear to have given transportation funding a great deal of thought, and aside from the gas tax, they were not very knowledgeable about the types of taxes or fees that fund transportation. Four of the eight groups knew that vehicle registration and licensing fees fund transportation, and not surprisingly, these tended to be the business and shipper groups. None of the groups mentioned their personal property taxes (applicable only in Northern Virginia), nor was anyone aware that a small portion of their real property taxes are used to fund local transportation.

In three out of the four general public groups, tolls were cited as a source of transportation funding, and one group also mentioned speeding tickets.

Groups don't have strong opinions on the current tax-based system for funding transportation; but they have lots to say about the *way* funds are spent.

After posing the question on how transportation was funded, the groups were presented with a list of the sources for transportation funding in their state. The list identified the source of the funding (i.e., the gas tax, a small portion of your real property tax, a small portion of the sales tax, etc.) without providing any budget figures.[3] Some respondents were surprised to see certain sources on the list (for example, one respondent didn't realize that a portion of his real property taxes funded transportation), and a few were curious about the amount of funding provided by the different sources. One Philadelphia resident said that he was dissatisfied with the current system, because "even if I don't use some of these services, I'm still paying for them." However, for the most part, respondents lacked interest and strong feelings regarding the current tax-based system for funding transportation; they tended simply to accept the system without thinking about it much.

In contrast, respondents had much stronger feelings about the way in which the funds are being spent, with many questioning whether the government effectively uses the funds it collects. In nearly all the general public groups, as well as the shipper and business groups, respondents mentioned that the government was not doing a sufficient job with their tax dollars. As one Northern Virginia resident put it, "I don't think they've done a very good job for all the taxes they get." This sentiment was particularly strong in

---

[2] More specifically, a Philadelphia business owner was frustrated by the DOT regulation setting 27,000 miles as the requirement for a commercial driver's license (CDL).

[3] The purpose of this exercise was to obtain respondent's impressions of the current tax-based system; attaching budget figures would have sidetracked the discussion into whether too much or too little is being collected by the different revenue sources.

Philadelphia, where all the groups complained about the poor quality of the roads and pointed out that Pennsylvania highways are among the worst in the nation (and roads elsewhere in the east coast are in significantly better shape).

- "I think we should have a lot less potholes in the road [given the taxes we pay]." (public, Philadelphia)
- "Pennsylvania highways aren't that great…the New Jersey gas tax is lower and their roads are nicer." (public, Philadelphia)
- "Out of the federal tax dollars, how much is really going back to roads? I don't know where it goes." (shipper, Philadelphia)

Congestion-based pricing: they get it, for the most part.

After the initial discussion on transportation issues and sources of transportation funding, respondents in each of the focus groups were presented with a scenario that described the congestion-based pricing system. After reading the scenario, they discussed their general impressions.

In each of the groups, respondents appeared to understand the general concept being described in the scenario. For instance:

- "The point is to get people to stagger their hours; they're going to get a benefit. Everyone won't be on the road at the same time." (public, Philadelphia)
- "If you are charging people to travel enough money, the roads will be used less – that's a fair economic point." (public, Northern Virginia)
- "[They are] trying to get cars off the road. This is one way to do it. People will try to rearrange their schedules…spread out the rush hour." (business, Philadelphia)
- "There are going to be winners and losers in the system." (public, Northern Virginia)

The only aspect of the new system that some respondents did not seem to understand was that they would no longer be paying the portion of their taxes that fund transportation. Interestingly, those in the group who *did* grasp this concept (that their transportation taxes would be eliminated) tried to clarify this point for those who were not so sure. Also, as detailed later in the report, some respondents who read the part about taxes being eliminated simply did not believe that the government would actually follow through with it.

What's the purpose?

While most respondents viewed the new system as a means for decreasing congestion, others (in several of the groups) had questions about its ultimate purpose. The following examples highlight the nature of respondents' confusion:

- "Is this supposed to be revenue neutral?" (shipper, Northern Virginia)

- "What do they want with this?" (shipper, Northern Virginia)
- "What is the purpose…fund transportation or keep people off the roads?" (business, Northern Virginia)
- "Are they looking to make more money?" (public, Philadelphia)
- "Can't figure it out…they're changing the whole way of getting money for it [transportation], so it's not just about traffic; it's got to be about something else." (public, Philadelphia)

A Philadelphia shipper viewed the main purpose of the system to be revenue generation: "To me, I thought this was about generating revenue so that the funding could be done differently rather than depending on government funding so much." This respondent felt this was a good thing; with increased revenue, "There is less reason to say things can't be done. The government is on the hook to produce results."

Another Philadelphia business owner was confused about the proposal, in light of the recent media publicity about the sale of the Pennsylvania Turnpike. "The Governor … is trying to sell the Pennsylvania Turnpike and you guys are trying to add more toll roads. What are we trying to do here? Do we have a plan? …It doesn't add up to me." In his view, if the government could not adequately administer the Pennsylvania Turnpike, then it should not be implementing new toll roads. Another respondent in the same group voiced suspicions about privatization, and feared that privatization of the turnpike would result in increased costs to drivers.

Congestion-based pricing: a mixed reaction.

Overall, reaction to the congestion-based system was mixed. When asked for their opinion, respondents with negative impressions tended to be the first to respond with their concerns and worries. A resident of Northern Virginia felt that a user-based system was unfair, and made the argument that transportation, like education, is a collective good: "I don't have kids in the school system; therefore should I not pay that portion of tax for education?" Nonetheless, a number of respondents (particularly in Northern Virginia) were not opposed to a "user-based" system, and even openly supported the idea.

- "[I] like the idea of pay per use." (public, Northern Virginia)

- "The idea of people paying for using something is not a bad idea. I don't have a problem with that." (public, Northern Virginia)

- "It's a use tax. If you use it a lot you should be paying for the use of it…More and more we're becoming a use society anyway. You pay for what you use." (public, Philadelphia)

The public's willingness to consider a congestion based system may be due, in part, to their sense that "something has to be done." Respondents in both cities felt that something drastic was needed in order to relieve traffic congestion.

- "People have to change their way of looking at it." (public, Northern Virginia)

- "Something has to be done, so if this is what they came up with, we have to let them give it a try." (public, Philadelphia)

- "We have to get outside our paradigm and come up with something different.... [this] needs to be looked at more…" (shipper, Northern Virginia)

- "Worth a test. Do research and see if it has worked there." (business, Philadelphia)

For some, their knowledge and/or experience with congestion pricing in other countries appears to have made them more open to the idea. In three of the four general public groups, respondents (without prompting) described the successes of congestion pricing systems adopted elsewhere. In Northern Virginia, for example, a resident who travels to London on a regular basis indicated that cordon pricing "works like a dream." Similarly, a Philadelphia resident described her experience on Toronto's tolled highway (Highway 407) as very positive.

People want more information on the program; in particular they want to know what their tax savings would be and how much they would be paying in tolls.

Through the course of the discussion, respondents had a number of questions on the mechanics of the new program and how it would work. They noted the need for additional information in order to assess the proposal. Some of their questions included the following (also see Appendix D):

- Would federal gas taxes be eliminated across the entire country, or just in major metropolitan areas? What about state and city gas taxes?
- What areas constitute a "major metropolitan area"?
- How is peak or rush hour defined?
- Would everyone have to get a transponder?

After reading the scenario, a number of respondents indicated that they had trouble determining their position on the new system because they were not sure what they were paying in taxes to fund transportation (i.e., what their tax savings would be), nor were they sure of how much they would be spending in tolls. Respondents wanted more information, so they could calculate their savings and costs. For many, this calculation was an important part of both understanding and assessing the new system.

After sharing initial impressions on the congestion-based system, respondents in the general public groups were given the opportunity to calculate their tax savings and their toll costs using worksheets developed specifically for focus group site. Respondents did not seem to mind the math exercise, and in fact, once respondents were able to think about what the system would mean for them, comprehension seemed to increase.

Interestingly, opinions of the system did not seem to necessarily correlate with who the winners and the losers were. Those who would be paying less still had concerns about the system, and some respondents who would be paying more said that the trade-off would be worth it, if indeed the roads were uncongested.

Congestion-based pricing system raises a variety of concerns.

In each group, respondents spent a significant amount of time discussing their concerns and worries about the congestion-based pricing system. Many of the same concerns were raised across the three types of groups: the general public, shippers and business groups; however, the respondents in Northern Virginia seemed somewhat more analytical and precise in expressing their concerns. Also, it is important to note, that with just a few exceptions, respondents were not angry or hostile about the likely ways they thought the system might fail. Instead, they seemed more dispirited or resigned, knowing from experience that plans to reduce congestion ultimately don't work out as originally advertised. They know, too, that some of the responsibility for reducing congestion falls to them, the drivers, and that they have been reluctant to change their own driving habits.

The respondents' concerns are summarized below, ordered roughly by frequency of response. Concerns that are specific to a particular city or type of group are highlighted accordingly.

Diversion to arterials: In all the groups, a key concern was that local roads would become extremely more congested. While respondents in each group noted that some drivers, particularly those with high disposable incomes, would pay the toll to drive on the highway, they still felt that many drivers would seek alternate routes in order to avoid the toll.

- "It will force all the traffic onto route 50 and congest the hell out of that… [it will] screw things up for local traffic." (public, Northern Virginia)
- "[This will] force you onto back roads. It will take you twice as long." (public, Philadelphia)
- "[There will be] congestion on smaller roads that might parallel these roads…neighborhoods up in arms." (public, Philadelphia)

A Philadelphia business owner understood that the diversion to arterials would be counterbalanced, at least in part, by a diversion of some to the highways. She reasoned that if people diverted to back roads, then it would make sense for her to take the highway to work. "If I drive to work I go the back roads…but if this happens more people are going to be on the back roads; maybe it won't work for me to go on the back roads."

Administrative costs: A serious concern raised in nearly all the groups was the administrative costs of implementing the toll system. Respondents perceived the system would require a large amount of start-up costs to outfit the highways, and to equip drivers with transponders. Respondents also felt that video-tolling, specifically tracking down drivers who do not have a transponder, would result in significant expense. This was a particular concern regarding visitors to the state. Nearly all the groups cited the need to ensure that visitors to the area pay their fair share of the tolls.

- "Will need an additional tax to cover administration [of this system]…costs you more to bill people than you will get back…" (public, Northern Virginia)
- "If photographing every single car, the costs will be huge…someone has to process." (public, Philadelphia)
- "Instead of building better roads, we're just going to put more people to work billing and checking transponders." (shipper, Northern Virginia)
- "What will it cost to run this? Will need to create an agency…" (shipper, Philadelphia)

Regarding administration, one Philadelphia shipper felt that the government would have to deal with a large number of complaints about toll bills. According to this respondent, average citizens would be confused about the rates and the timing of different rates, and would frequently question the amount of their bills. This same respondent also mentioned that many people in Philadelphia don't have credit, and he wondered how the system would be administered among that population.

Distrust of government: In both Northern Virginia and Philadelphia, there was a perceived lack of trust in government to effectively administer the new program. For example, despite the information presented in the scenario, many respondents voiced doubts that the existing taxes that fund transportation would really be eliminated.

- "Don't believe taxes would go away…don't have confidence in it for that reason." (public, Northern Virginia)
- "I don't think we're going to get a reduction in taxes." (public, Northern Virginia)
- "They'll still keep our taxes." (public, Philadelphia)

Respondents also voiced concern about accountability, and questioned whether or not the government would use the funds collected from tolls for transportation improvements.

- "Do we believe the government, that the money will go back into transportation?" (public, Northern Virginia)
- "This is another boondoggle; the money won't go to public transportation." (business, Philadelphia)
- "You're still sending it [toll revenues] to some kind of government or political body to spend money…What's going to make whoever is in charge of this money spend it on the most appropriate places to spend it? You can't do it." (business, Northern Virginia)

For others, particularly in Philadelphia, a key worry was the government would simply keep raising the price of the toll. As one respondent put it, "Once we get a tax passed, it's real easy to increase it and difficult to take it away." Others expressed similar sentiments:

- "The biggest issue is a trust in the government issue people have. Is 25 cents going to turn into 35 cents, 50 cents…" (business, Philadelphia)
- "Rates adjusted four times a year – [that is] scary. If [the system] didn't do well last quarter, bump it up." (public, Philadelphia)
- "Prices have gone up on the Greenway…the same will happen with toll roads." (business, Northern Virginia)
- "They are going to rate it four times a year...eventually just keep increasing rates." (shipper, Philadelphia)

A Philadelphia business owner worried that people would not notice increases in their tolls, since tolls would be automatically deducted from their E-ZPass accounts.

In related comments, respondents conveyed that they have been let down before, and they fear that this is just another 'scheme' or plan that won't work.

- "There have been so many schemes that have come along in this area tied to transportation. Every time it doesn't work out the way we thought it would." (business, Northern Virginia)
- "Mark Warner went all over Virginia…got everyone to vote to raise taxes…we never saw a dime." (business, Northern Virginia)
- "We get promised a lot and everything never happens." (business, Philadelphia)
- "Need to convince me that they are going to give me a better product… [I] want a sign that things are better." (public, Philadelphia)

High personal (business) cost: A number of respondents in both cities and across the different types of groups noted that the system would be a financial burden and estimated that it would cost them more than they are currently paying in taxes. This reaction to the new system was particularly strong in Philadelphia.

- "Costs [are] too extreme for everyday driver." (public, Philadelphia)
- "It's not just the cost of the toll, but others passing their costs onto you." (public, Philadelphia)
- "Will hit people hard…[they] will pay more in tolls than in taxes." (business, Northern Virginia)
- "It's a lot of money. Will cost me more." (public, Northern Virginia)

One Philadelphia respondent indicated that she did not like the new system, in part, because she thought she would "feel" the tolls more, whereas the current taxes are, in a sense, "hidden."

Business owners voiced significant concern about the increased cost to their business. For many businesses, the location of their office and/or the nature of their business meant that employees were constantly on the highway, either commuting to or from work or visiting client sites. A business owner in Northern Virginia, for example, was concerned that his employees would seek higher wages to cover the costs of the new tolls. In

Philadelphia, a business owner worried that people would move out of the city as a result of the new system, thus hurting the city's economy. Another respondent in the same group, however, suggested that people might move *into* the city (these opposing comments indicate there is some confusion about the land development impacts of congestion pricing).

Compared to the business groups, the shipping groups did not seem as concerned about the costs. Respondents noted that they will simply pass the costs onto their customers, and some indicated that their commercial fleets drive at night, so they would not be paying the toll.

Income equity issues: The Northern Virginia general public groups, which tended to be more affluent (compared to those in Philadelphia) demonstrated greater concern about the ability of low income people to pay the tolls. While it was not a predominant concern, a number of respondents raised the equity issue on their own, expressing that those hit hardest will be those with the lowest incomes, as their schedules are less flexible, they live further out due to housing prices, and the peak toll would eat up approximately one hour's worth of wages. Income equity issues were also raised in the Philadelphia business and shipper groups, though the topic did not generate a strong reaction among participants.

- "People who are less affluent are going to have problems with it." (public, Northern Virginia)
- "Use tax is regressive…kills people." (business, Philadelphia)
- "For some this would be an economic hardship. [They] already have trouble paying their bills." (shipper, Philadelphia)

In the Philadelphia general public groups, where the respondents were relatively less affluent, they tended to personalize the issue of equity. That is, they focused on the high costs of the system to themselves.

When respondents were asked how they felt about low-income people receiving a toll discount, reaction was mixed. While most tended to think it was a good idea in theory, there were concerns about the criteria that would be used to qualify low income drivers for the discount, as well as concerns about administering such a toll discount. Many felt that it would increase the overall costs of the system and that it would provide an avenue for those who wanted to cheat the system. For example, a respondent in both Northern Virginia and Philadelphia wondered, "What would stop me from driving my grandmother's car?"

Won't solve the problem: A number of respondents simply could not fathom that highways would be free of congestion; they had difficulty understanding and accepting this concept. Despite admission by some that they would consider changing their schedule, mode, or route if the system were implemented, many respondents did not seem to make the connection that these changes in behavior would result in less congestion on the highways.[4]

- "How do you know if the roads will really be less congested? It's hard to imagine until the day is really there." (public, Northern Virginia)
- "I don't know how this is going to stop rush hour traffic. It is still going to be congested. It doesn't solve the pure math of number of cars on the roads." (public, Philadelphia)
- "I don't think enough people will stop driving the roads." (public, Philadelphia)
- "How can all roads be less congested?" (business, Northern Virginia)
- "As long as people can buy gas, the roads will remain congested. People will get in their cars and drive." (shipper, Philadelphia)

Other respondents were unsure if people would really change their behavior:

- "I think it's hard to predict what's really going to happen. They don't know how many people are willing to change." (public, Northern Virginia)

In Northern Virginia, a couple of respondents stated that by the time this new system were implemented, there would be significantly more drivers on the road, making it less likely that the system would work.

In addition, a Philadelphia resident noted that with one accident, traffic would be back to stop-and-go.

Enforcement difficulties: A number of respondents expressed doubt that the government could effectively enforce the system. This was a particular concern regarding visitors to the area and others who do not have a transponder. In addition, respondents felt that people will try to "cheat" the system.

- "People will try to fly through and get away with it [not paying the toll]." (public, Northern Virginia)
- "A lot people running it, not concerned about the consequences…people owe thousands of dollars in parking tickets." (shipper, Philadelphia)

As noted earlier, cheating was also mentioned as a concern in response to the possibility of a toll discount for low-income or seniors.

Privacy: Privacy was raised in nearly all of the focus groups, however, it was not an issue that resonated strongly or generated much discussion. As one Philadelphia resident stated, "privacy is almost gone anyway." Nonetheless, for two respondents, this was an important issue. A Philadelphia business owner (mentioning national new stories about

---

[4] Focus groups conducted in Washington State found similar results; participants did not seem to grasp how converting an existing lane into a HOT lane would speed traffic. According to the report, "pricing for traffic management is not a concept near the tops of their minds." (Lawrence Research, " A Two-Phase Study of Attitudes Of Washington State Voters Toward Transportation Issues," April 2006).

warrantless searches and eavesdropping) repeatedly raised concerns about privacy and wanted to know what would be done with the data that was collected. A shipper in Northern Virginia was worried that "This is big brother." Other respondents in those groups, however, did not follow their lead on this topic.

Restraint of choice: This issue was raised in a number of groups. In Northern Virginia, for example, respondents felt their choices were being limited, and they expressed dissatisfaction that on the tolled highways they would not be able to choose whether or not to pay the toll. As one respondent explained, "Having a choice is key. Something like the Dulles Toll Road, you have a choice whether you want to pay or whether you don't, going to the same location." Another resident referred explicitly to High Occupancy Toll (HOT) lanes: "What about choice? Optional HOT lanes on 66? If you want to pay to go faster, go ahead."

Others spoke of the need for improved public transportation options, so that they could choose whether to pay to use the roads or whether to pay to use transit. In their view, since the new system is trying to get people off the roads, then it is only fair that the public should have viable public transportation options. As one Philadelphia resident stated, "If public transportation is improved, then you have a choice. Otherwise [the toll is a] forced cost." In other words, these respondents felt that improved public transit (particularly expanding service) should be an integral component of the congestion pricing system.

In related comments, groups in Northern Virginia (but not Philadelphia) referred to congestion pricing as un-American, a sentiment that seemed to stem, in part, from their complaint that tolling restrained their choices.

- "It's just another thing in America you're losing your freedom to do whatever you want when you want." (public, Northern Virginia)
- "Used to be the American way – I as an American can go anywhere in my country…." (business, Northern Virginia)
- "Something un-American about not having a free road." (public, Northern Virginia)

"Commuter" equity issues: Respondents felt it was unfair that commuters should bear the brunt of the tolls. A Philadelphia shipper, for example, claimed that the system would be "taxing the wrong people." In his view, people who have to get to work every day will have no choice but to pay the toll, whereas large trucking companies will be able to alter their driving routes or times to avoid paying tolls. He pointed out that these companies use lots of fuel and will pay neither gas taxes nor tolls. A number of respondents in both cities also voiced concern that those with less flexible schedules would be paying more.

- "People that have to travel the roads when they do carry the weight of others." (shipper, Philadelphia)

- "Some people can change time [they travel]; I don't have that option." (business, Philadelphia)
- "People on the road during peak have to be there for their job." (public, Northern Virginia)

A business owner in Philadelphia also felt that it was unfair that people with longer commutes would have to pay more. Several Philadelphia residents suggested that in addition to charging commuters tolls during the week, tolls also should be paid by leisure drivers on weekends, since congestion is a problem on weekends as well as weekdays.

In Northern Virginia, where respondents were more actively engaged in analyzing the different implications of the new concept, they offered the following concerns or worries that were not explicitly mentioned in Philadelphia:

Toll revenues insufficient: With the elimination of transportation taxes, some respondents questioned whether the toll revenues would be sufficient to fund all roads in the area.
- "How can you take away all this funding and only charge for driving on freeways – there are lots and lots of other roads…." (public, Northern Virginia)
- "Don't see how you can remove all that [taxes] and still have enough." (public, Northern Virginia)

Unfair to highway drivers: There was also some concern, raised in Northern Virginia, about the fairness of highway drivers having to fund <u>all</u> the roads – highways as well as local roads. As one respondent asked, "Why should people who pay for highways [through tolls] also be paying for maintaining [route] 50?" In a related comment, a Northern Virginia resident who tends to drive mainly on local roads felt it was unfair that she would neither be paying tolls nor transportation taxes under the new system, despite using the roads.

System is punishment: Respondents in the Northern Virginia groups referred to the tolls as a "punishment" or "penalty," a sentiment that was not voiced in Philadelphia.

- "I can see the benefit of charging commuters money for using roads, but I feel like I'm being punished for wanting to go somewhere because the way we are doing it now is not sufficient." (public, Northern Virginia)
- "The money is being put into techniques that are designed to force behavior and punish people who don't fall into the category of having the money to be able to go faster or having the number of people in their car that allows them to get into the fast lane." (business, Northern Virginia)

This reaction may be due to contextual factors in Northern Virginia. Contemporaneous with the focus groups, the media had been covering the implementation of a new plan to significantly increase fines for traffic violations to fund infrastructure. Respondents in all the groups spontaneously raised this issue at some point during the discussion and expressed a strong negative reaction to the plan.

In addition, a shipper in Northern Virginia who described the system as a penalty felt that the government and business are constantly at odds: "You're talking about entrepreneurs and government and those are cats and dogs. I don't think you can get either one to see the other person's perspective."

Issues raised specifically in Philadelphia (but not Northern Virginia) included:

Creation of new rush hour: In Philadelphia a few respondents (general public and business) suggested that the toll system would simply create a new rush hour, without necessarily easing traffic congestion. For example, if the toll were charged from 4 p.m. to 7 p.m., then a new rush hour would start at 7 p.m. as drivers seek to avoid paying the toll.

Seniors opposed: In one of the Philadelphia general public groups, a few respondents voiced concern that seniors would be opposed to the system. In their view, seniors are afraid of change and will worry that their credit card is being billed in error. Interestingly, the few people in their sixties who participated in the groups did not mention this concern themselves.

Geographical equity: A representative from a shipping firm in Philadelphia stressed that the system has to be uniform across the entire country: "if you paid [gas] tax in New Jersey, then get on the road and are tolled, you will be double hit." Similarly in the general public focus group, a Philadelphia resident said it should be "nationwide…all or nothing."

Overall, the business and shipper groups had the same concerns as the general public; however, they also mentioned these two additional issues:

Tough to budget for businesses: A shipper in Northern Virginia felt that it would be difficult for businesses to budget for the new system. In both the shipper and business groups in Northern Virginia, respondents suggested charging businesses a flat rate.

Jurisdictional disputes (Federal vs. State): A shipper in Northern Virginia raised the concern that states will balk at the U.S. DOT plan and will view it as "taking away our rights."

In an "alternate universe," the public perceives the benefits of congestion pricing.

There were only a few respondents who expressed positive opinions about the system from the start. A Philadelphia resident, for example, claimed "To speed things up I would pay anything to be able to drive one speed to work…" A biking enthusiast from Northern Virginia felt "It might work…it's incentive to carpool, then there would be less traffic… For every person that it doesn't work for, there are ten that jump to the idea. We're talking about transportation flow and congestion – somebody's got to pay for it."

In most of the groups, however, respondents seemed to get 'stuck' in their discussions, circling around the notion that either the government would not follow through on the new system or that the roads would remain congested. When asked to imagine that they lived in an alternate universe where everything worked as it was supposed to, respondents were able to more clearly articulate the benefits provided by the system, including reduced congestion, time savings, and better roads. One respondent also offered "replenished ozone" as a benefit of congestion pricing. A Northern Virginia resident noted that "The trade-off in quality would be worth it for me. If commutes were faster, there would a quality rise in my traveling and given the relative trade-off in cost, that would be worth it."

For business and shipping firms, respondents indicated that their employees could get to work faster, there would be less idle time, reduced fuel consumption and possibly a reduced workforce. One Northern Virginia business owner suggested that with decreased traffic congestion she might have greater access to a higher-paid employees market (i.e., employees who previously would not take the job due to the commute might now consider it). At the same time she noted that her access to a lower-paid employee market might worsen as a result of the tolls.

Another business owner in Northern Virginia felt, "If there were no traffic congestion, [it would be an] immense, huge paradigm-shifting increase in productivity." A Philadelphia shipper who supported the concept compared toll rates to cell phone rates (both of which are variable), and noted that as the cell phone industry matured, rates have decreased. In his view, this "would be extremely palatable."

Some would change their travel behavior, or where they live and work.

A number of respondents across the two cities indicated that the new system would cause them to reevaluate their current driving behavior, as well as where they live and work. A Northern Virginia resident with a long commute claimed that if her employer did not assist her with the cost of tolls, she would have to look for another job closer to home. A currently unemployed woman (Northern Virginia) said she would only consider jobs that did not require a long commute, and a resident of Philadelphia indicated that with the new system she would try to find a job closer to home.

However, a Northern Virginia respondent who has been commuting 22 miles to the exurbs for 25 years said she would not have the flexibility to switch jobs, as she is very close to retirement. She did not view moving as an option either, since she could not afford a comparably-sized house closer to her place of employment.

A number of respondents said that they would switch to public transportation, if service were expanded to their area. For example, a Philadelphia resident described how he did not own a car when living in New York and relied solely on public transportation, but currently he did not have convenient access to public transportation, so he had to use his car.

Telecommuting, flextime, carpools, and express bus service have some appeal.

In combination with tolling highways, the scenario described several additional components of a user-based congestion pricing system. For example, express bus service would be developed, and employers would be encouraged to provide their employees with telecommute and flextime options. In addition, certified vanpools would be exempt from the tolls and carpools would be able to split the cost of the toll among all the riders. Across all the groups, response to these components of a congestion pricing system was generally favorable; however, a number of respondents noted that their work situation did not allow for telecommuting or flextime. In northern Virginia, a government employee and a government contractor explained that they need to be in the office during "core" hours, so "there is a limit to flextime." In addition, business owners in both Northern Virginia and Philadelphia, particularly those in retail and consulting, felt that telecommuting is not really an option; retail stores have fixed hours of operation when their employees are required to be at the store, and similarly, consultants explained that they have to be in the office "to service their clients." Moreover, several respondents – particularly business owners – noted that many people are already using flextime and it "isn't working" (to ease congestion).

With regard to carpooling, several respondents were already carpooling, and indicated that their arrangement worked well. A few respondents in both Northern Virginia and Philadelphia noted that congestion pricing was an "incentive to carpool," and that some people would take advantage of this option. However, many seemed reluctant to join a carpool and preferred the convenience of driving their own vehicle. That is, carpooling was perceived as a good idea for other people.

Respondents were not overly enthusiastic about using express bus service, in part because they needed more information about its convenience and reliability before they would consider using it.

E-ZPass is popular.

A number of respondents in all the groups said they have an E-ZPass, and those who do not are aware of the technology. Overall, the response to E-ZPass was very favorable, and a number of respondents described how E-ZPass has saved them time in their travels. As one business owner explained, "E-ZPass seems to have done wonders."

Regarding the new system, there was a mix of opinion on whether all drivers should have a transponder. A couple of respondents in the Philadelphia general public groups were concerned that people would be "forced" to get a transponder. Others did not share this concern, and felt that for the system to work all drivers *should* have a transponder.

A Philadelphia respondent also thought it was important for people to know exactly when and where they were being tolled.

Environmental benefits not on most people's radar, though potential benefits acknowledged.

In nearly all the groups, respondents did not mention the potential environmental benefits of the congestion pricing system. When asked about the issue, they acknowledged that there might be benefits, with several mentioning reduced emissions. A Philadelphia business owner pointed out that carpooling or switching to public transportation would help the environment. However, he qualified his statement by saying that he was not sure such changes in behavior would take place. Another respondent in the group felt that "No more stop and go" would be "huge" in helping the environment. One respondent advocated for getting rid of cars that only get 15 miles to the gallon ("SUVs"), but his reasoning was not based on environmental benefits; rather, he explained, "To me it's an efficiency issue. We need to be more efficient overall." In general, the discussion on environmental benefits did not generate much reaction or interest in the group discussions.

Consider these improvements…

In the course of the discussions, as respondents thought about congestion pricing and how it would affect them, they offered a number of ideas for ways to make the overall concept more palatable, including suggestions for how the system should be marketed. Their comments are summarized below:

Provide toll exemptions for owners of hybrid vehicles. Several respondents (business groups and general public) recommended that owners of hybrid vehicles, such as the Prius, should receive an exemption or a toll discount. One Philadelphia business owner pointed out that Prius owners would not do so well under the new system (compared to a Lexus owner), because they would benefit less from the gas tax exemption (their cars get such good gas mileage that they pay less in gas tax). He proposed charging Hummers a significantly higher toll compared to Prius owners ($1 per mile versus 10 cents per mile).

Incentivize the purchase of E-ZPass. In both general public groups in Northern Virginia, a respondent suggested giving toll discounts to E-ZPass owners, as this would create an incentive to acquire the technology. The respondent saw this as a means for keeping administrative costs down; if more drivers have the E-ZPass, there will be less need to "track down" those without a transponder.

Provide a tax refund to drivers at the end of the year. A resident in Philadelphia was skeptical that his taxes would decrease if the congestion pricing system were implemented. He wanted something tangible – a reward – that would confirm his tax savings. In a related comment, a Northern Virginia business owner suggested "flipping" the system by rewarding people who don't use the roads during peak (rather than "punishing" people with tolls).

Toll proportional to wear and tear imposed on the highway. A Philadelphia business owner suggested that tolls should vary according to the size of the vehicle (similar to the way in which different sized trucks pay different tolls on the New Jersey Turnpike). This would encourage the public to buy smaller cars.

Charge commercial companies a flat rate. Representatives from business and shipping in Northern Virginia recommended that the commercial sector pay a flat tax (i.e., per vehicle per year) for use of the tolled highways. This would enable firms to better budget for the new tolls; otherwise, they envision their costs being astronomical.

Be up-front about the costs of the new system. Several Northern Virginia residents felt that the congestion pricing scenario presented to them painted too rosy a picture and was not up-front about the costs of the new system. As one resident explained, "I'd be more apt to sign on to a situation like this if it wasn't sold to me as a neutral thing in terms of my pocketbook. Unless it's sold relatively honestly, where we say it's going to cost and it's going to force changes in behavior…"

Communicate the benefits. "To sell this idea" a Philadelphia business owner felt it was important to communicate exactly how much money in taxes would be eliminated under the new system. "You've got to make it very clear that out of $3.02 per gallon that $1.62 (or whatever it is) is taxes that will be eliminated."

Provide a guarantee of no congestion. One Philadelphia business owner suggested that drivers should not be charged if the system does not work (i.e., if there is traffic congestion), an idea supported by others in the group.

Eliminate gas taxes before implementing road tolls. A Philadelphia shipper indicated that the public would embrace the system if taxes at the gas pump were eliminated prior to implementation of the tolls. Perhaps he saw this as a means for increasing public confidence that the government would follow through on eliminating gas taxes.

Charge out of state drivers more. A shipper in Northern Virginia felt that out of state drivers should be paying more to fund the roads in Northern Virginia. Even when someone pointed out that this meant he would pay more to drive through other states, he still advocated for higher fees for out-of-state drivers.

There are other ways to fix the problem.

Respondents in each of the focus groups periodically offered alternative ideas to congestion pricing as a means of solving traffic congestion. In many cases, respondents indicated that the congestion-based system was "too complex," and they felt there were simpler options for solving the problem.

Improve public transportation. After reading and discussing the scenario, a number of participants across both cities returned to the idea that the best way to decrease congestion is to improve public transportation (in many groups, and especially in Philadelphia, better transit service was offered as a solution to traffic congestion in the opening discussion). If rail service were expanded to outlying areas, and if the service were more reliable, many respondents said they would take public transportation. As one Philadelphia business owner claimed, "Build the rails and they will come." A Philadelphia shipping firm respondent suggested educating younger people to use public transportation, since older people are not likely to change their habits. At the same time, he indicated that public transportation needs to be made more enticing: "Clean it up and make it safe."

Other ideas for solving the congestion problem were offered by just one or two respondents:

Raise the gas tax. A respondent in Northern Virginia as well as Philadelphia mentioned this as a simpler alternative to the congestion based system presented in the scenario. Their comments also suggested that it was more equitable. As the Northern Virginia resident stated, "it's going to cost more to get people off the roads, and the best way to do it is the gas tax…if you raise the gas tax it will take care of the whole thing. People who drive more will pay more in taxes." (public, Northern Virginia).

Bill people on the size of their vehicle's engine and annual number of miles driven. A northern Virginia resident offered this idea as a simpler alternative to congestion based pricing, and one that would encourage the use of smaller cars.

Fund a Federal initiative on energy. One Philadelphia business owner suggested that the government should conduct research on alternative fuel sources, increased mileage and alternative forms of transportation. The government needs to make a commitment to energy "in the same way as when Kennedy announced the space program."

Conclusions and Recommendations

This research effort was undertaken to obtain a better understanding of the public's views toward a new congestion pricing scenario and to learn how to best communicate with the public about this topic. Focus groups were conducted in Northern Virginia and Philadelphia with the general public, business groups and shipper groups to explore these issues in greater depth. Based on the findings, key recommendations are highlighted below.

The public needs to be better educated on how transportation is funded today.
While many of the respondents knew that the gas tax funds transportation, they were generally unaware of the amount of the gas tax and few could name any of the other sources of transportation funding. In order for the public to comprehend this congestion pricing concept, they need to have a better understanding of how transportation is funded today. Equipped with this knowledge, they will be better able to assess the implications of the new system.

Be clear about the purpose of congestion pricing initiative.
The public will be more willing to consider this concept (and possibly support it) if they understand its purpose. The scenario for this study represents a relatively novel approach for funding transportation and for dealing with traffic congestion. Given this lack of familiarity with the concept, it is especially important to explain why this particular approach is being adopted. In the focus groups, some respondents were not quite sure about the purpose of the congestion pricing initiative (was it congestion mitigation? revenue generation?), and this made them hesitant to support the concept. Respondents also mentioned that there were simpler ways, such as raising the gas tax, to achieve the revenue generation objective. Hence, any initiative should explain why a new system is needed, and specifically, why this new system is better than raising the gas tax.

Explain how the congestion pricing initiative will affect individuals and households.
Early in their discussions (before completing the worksheets on tax savings and toll costs), focus group respondents indicated that they needed to know what their tax savings would be and what their toll costs would be in order to assess the new concept. In other words, they wanted to know how the system would affect them personally. And after completing their worksheets, respondents seemed to have a more concrete understanding of the system and what it would mean to them personally. Thus, when communicating to the public about this concept, it is important to convey which taxes would be eliminated, and how much would be saved as a result. In addition, respondents need to be made aware of the costs of the new system to them.

Acknowledge both the costs and the benefits of the new congestion pricing system.
As one respondent advised, the concept should be presented in an honest way, acknowledging both the benefits and the costs. Painting too rosy a picture will only provoke cynicism. As one respondent explained, "I'd be more apt to sign on to a

situation like this if it wasn't sold as a neutral thing in terms of my pocketbook…[it] needs to be sold honestly."

Educate the public on the magnitude of travel behavior shifts required to achieve free flow travel conditions.
The public's difficulty in grasping (or believing) that congestion pricing will create free flow travel conditions is partly due to the fact that they seemed to have no sense of how much shift in travel behavior would be necessary to insure uncongested conditions. That is, how many people would have to switch to telecommuting, flextime, carpooling, or use the express bus service in order to ease traffic congestion? Findings from the focus groups suggest that people may be unaware that relatively small shifts in behavior are required to produce free-flow conditions.

Clearly explain how the toll revenues will be used.
Across the groups, respondents had questions about how the toll revenues would be used, and they voiced doubts that the funds would be invested in transportation. To allay these concerns, the public should be given specific information on how the toll revenues will be spent; moreover, it is critical that they see tangible improvements in those areas.

Explain how you will meet the traffic diversion challenges posed by the congestion pricing concept.
Many respondents had serious concerns about the diversion of traffic to arterials and felt this redistribution of demand would have a significant, negative effect on their driving experience. This issue should be addressed up-front to allay public concern.

Attend to the public's concerns about the administrative costs of the system.
The administrative cost of implementing the congestion pricing system was raised repeatedly in both cities and across the different types of groups. Respondents perceived the system would require a large amount of start-up costs to outfit the highways, and to equip drivers with transponders. Respondents also felt that video-tolling, in particular tracking down drivers who do not have a transponder, would result in significant expense. The public needs more information on the costs of administering the program and how the costs will be handled.

Address issues of equity.
In Northern Virginia, where respondents tended to be somewhat more affluent and better educated (compared to Philadelphia), the issue of income equity was an important value and seemed to resonate more strongly. Respondents voiced concern that congestion pricing was regressive and would adversely affect lower income people, a segment of the population least able to absorb the additional costs of the tolls. In Philadelphia, the respondents were lower income, and they tended to focus more on the high cost of the program to themselves; a number indicated that congestion pricing would be a significant new expense. These findings suggest that issues of economic equity must be addressed with any congestion pricing initiative.

Explore privacy issues.
While this issue was a lesser concern relative to the others, it was mentioned in nearly all the groups, and there were a couple of respondents who felt very strongly about it. This finding suggests that further exploration of the issue is warranted. For example, this issue has been addressed by using "smart cards" with stored value in Singapore, and by providing the option of anonymous accounts in Orange County, California and Puerto Rico. The focus group findings indicate that a potential means of easing privacy concerns is to let the public know exactly what data is being collected and how that data will be used.

Consider ways to address the economic concerns of businesses.
Business owners seemed particularly worried about the costs of the congestion pricing system to their business (including their suppliers, customers and employees), and the new administrative costs that such a system would create. Any congestion pricing proposal should be sensitive to the concerns of the business community, especially small businesses, demonstrating that those concerns are being taken into account.

Consider the suggestions advocated by the public.
The focus groups had a number of suggestions for making the congestion pricing system more palatable. Future work should explore the appeal of these different proposals, including:
- Providing toll exemptions or discounts for owners of hybrid vehicles
- Incentivizing the purchase of E-ZPass
- Providing a tax refund to drivers at the end of the year (i.e., for their tax savings)
- Adjusting the amount of toll proportional to wear and tear imposed on the highway
- Charging shipping, logistics and fleet-based businesses a flat rate to reduce the administrative cost of doing business
- Providing a guarantee of no congestion (i.e., drivers do not have to pay the toll if the roads are congested)

Educate the public on congestion pricing systems in other countries.
Respondents who had personal experience with congestion pricing systems in other countries or had read about such initiatives seemed to be more comfortable with the concept. To the extent that the public becomes familiar with congestion pricing initiatives elsewhere (and the resulting benefits), the concept will seem more "believable" to them. Currently, the public has a hard time imagining that congestion pricing could truly create free flow traffic conditions. Indeed, a number of respondents, in search of evidence that the system could really work, asked whether the concept had been implemented elsewhere. This echoes findings from a focus group study that was conducted to explore the public's attitudes toward Intelligent Transportation Systems (ITS).[5] In those groups, the public repeatedly asked for information on the performance of ITS in other (similar) cities. Their support of future investment in ITS hinged on

---

[5] Petrella, Margaret and Jane Lappin. "Information Needs Assessment Activity: Results from the General Public Focus Groups," September 2006.

gathering more information on the costs and benefits of the technology, lessons learned and examples of ITS successfully deployed in other cities. This type of evidence will help to allay the ubiquitous concern that any new system, no matter how promising it seems, will eventually be ineffective in controlling congestion.

<u>Be sensitive to the socio-political context of the metropolitan areas where congestion pricing is being implemented.</u>
The political context in each of the cities had an impact on views about the congestion pricing initiative. In Northern Virginia, a plan under consideration that would significantly increase fines for Virginia residents seemed to influence their perception that the new tolls were a form of "punishment." In Philadelphia, recent media publicity about the sale of the Pennsylvania Turnpike had some respondents feeling uneasy about privatization and about adding more toll roads. These findings suggest that current political issues that might affect perceptions of congestion pricing need to be taken into account; messages to the public will need to be tailored based on the specific political context within the metropolitan area.

Appendix A: Discussion Guides

MODERATOR'S GUIDE: GENERAL PUBLIC

Congestion Pricing Groups, Fairfax, VA and Philadelphia 7/07
(95 minutes total)

I. Introduction (5 minutes)
Focus group ground rules & format
Respondents' self-introductions: name, occupation, typical travel in the area.
General topic of the discussion – transportation in this area

[NOTE: The following questions/subject areas may not be presented verbatim or in order. All may not be presented in every group and additional issues may be included. Their purpose is to stimulate thought. The discussion will flow naturally, with interventions from the moderator to keep the conversation on track and to probe specific subject areas.]

> The main objective is to assess the public's receptivity to changing from the current tax based transportation system to congestion based/user paid system.

II. Warm Up: Important Issues/Transportation Issues (10 minutes)
When you think about the transportation issues that you face in your everyday life, what comes to mind? *[Listen for/probe for traffic congestion. Follow up with:]*

What do they think can be done about traffic congestion? Is it solvable problem?

[Listen for/explore – but do not mention]: Gas prices i.e. how have high gas prices affected your travel? Pollution/environment]

III. Main Discussion *[building on conversation above]* (75 minutes total)
<u>1. Understanding of Transportation costs and how they are paid for</u> (Max 15 minutes)
Continuing with our discussion on transportation, what's your sense of how current transportation costs are paid for? i.e. how do we fund transportation? [**DO NOT PROMPT.** Be aware of concern re new VA taxes]

[IF GAS TAX MENTIONED] Do you know how much tax you pay on gas?

[Present list of transportation funding sources on easel pad. If they missed a number of items used to fund transportation]: Are you surprised by this list, by the ways in which transportation is funded?
What do you think of our current system for funding transportation?

2. <u>Taxes Vs Tolling Part I</u> –(25 minutes)
[Explain that these taxes have been insufficient to cover transportation costs, so another method is being considered.] [Present handout with general scenario – participants read]

- Discuss perceptions of the two mechanisms for financing transportation. What are your questions? What concerns you?

- What do you think of a system such as this where the driver pays per miles consumed on the highways – also known as a "pay as you go" system? [Probe: charges varying by level of demand, so that rates for miles traveled on some highways would be higher than others, and rates would be higher during the peak hour than in "shoulder" hours]
- Do you think there are any benefits to this new toll-based system versus a tax based system?
- Under what conditions would you accept a toll based system in place of a tax based system?

[Listen for/probe:
- [equity issues/double taxation] Does this system seem equitable or fair? [Probe:] What equity issues concern you? [Follow up with:] *What if there was a toll-discount for low income individuals who must make work-related trips during rush hour? What if low income individuals received free service, or no tolls?*]

- [Explore understanding that pricing can provide free flowing traffic] From what you read, how would this new road pricing system provide free-flowing traffic during peak travel times? Is this believable? [listen for/probe: what do you think if the toll rates had to be adjusted periodically to maintain free-flowing traffic]

- [ETC and privacy concerns] What do you think about the electronic tolling technology mentioned in the write-up? [if necessary, probe:] Do you have any issues or concerns? Do any of you have an EZ Pass or Smart Tag? How often do you use it? How does the system work for you?

- [pricing as an environmental strategy] From an environmental point of view, are there any benefits to this new road pricing system? How important are they?

3. Taxes vs. Tolling Part II – Additional Components (15 minutes)

The write-up described some additional features or services that would implemented along with the tolling. These included express bus service, more park-and-ride lots, vanpools and carpools, and flextime. [REFER PARTICIPANTS TO SCENARIO:]

[NOTE: Loudon County commuter bus service is $60 for 10 one way tickets, from Leesburg into DC/$15 for 10 one way trips from Leesburg to West Falls Church Metro]

Discuss: Perceptions of these additional features of the system, affect on overall opinion about the road pricing system described in the write-up, appeal of various components.

Likelihood of using the express bus service, carpooling, or taking advantage of flex time?

4. Costs vs. Savings of Scenario (20 minutes)

Now we would like you to think more concretely about what the new system would mean for you. Here is a map of this area that highlights the roads that would be tolled.

[PROVIDE MAP]
Let's look at some tables that provide information on the tax savings as well as the costs of this new system.
[Present tables with specific costs/savings. Give participants calculators/worksheets. Walk them through calculation. If relevant, they can perform the calculation for their spouse only (instead of themselves) OR for themselves and their spouse.]

Based on your calculation, what do you think? [Probe: Likelihood of changing travel patterns if this new road pricing system were instituted on the highways?]

IV. Conclusion (5 minutes)
Final comments, Summary, Thank You's, Good bye's.

MODERATOR'S GUIDE: BUSINESS GROUPS
Fairfax, VA and Philadelphia 7/07
(95 minutes total)

I. Introduction (5 minutes)
Focus group ground rules & format
Respondents' self-introductions: name, business
General topic of the discussion – transportation in this area

[NOTE: The following questions/subject areas may not be presented verbatim or in order. All may not be presented in every group and additional issues may be included. Their purpose is to stimulate thought. The discussion will flow naturally, with interventions from the moderator to keep the conversation on track and to probe specific subject areas.]

> The main objective is to assess the public's receptivity to changing from the current tax based transportation system to congestion based/user paid system.

II. Warm Up: Important Issues/Transportation Issues (10 minutes)
What role does transportation play in running your business? [Discuss components of business mediated by transportation i.e. suppliers, shipment of product, employees, customers]

What transportation issues impact your business? *[Listen for/probe for traffic congestion. Follow up with:]*

What do they think can be done about traffic congestion? Is it solvable problem?

[Listen for/explore – but do not mention]: Gas prices i.e. how have high gas prices affected your business? How have you responded to high gas prices? Pollution/environment]

III. Main Discussion *[building on conversation above]* (75 minutes total)
<u>1. Understanding of Transportation costs and how they are paid for</u> (Max 15 minutes)
Continuing with our discussion on transportation, what's your sense of how current transportation costs are paid for? [**DO NOT PROMPT. Be aware of concern re new VA taxes**]

[IF GAS TAX IS MENTIONED:] Can anyone tell me how much you pay in gas taxes? (i.e. cents/mile)

[Show list of different taxes that fund transportation]

What do you think of our current system for funding transportation? [PROBE: Do you see any problems with the current funding system?]

2. <u>Taxes Vs Tolling Part I –(40 minutes)</u>
[Explain that these taxes have been insufficient to cover transportation costs, so another method is being considered.] [Present handout with general scenario – participants read]

- Discuss perceptions of the two mechanisms for financing transportation. <u>From the perspective of your business</u>, what are your initial impressions of this new model for funding transportation, where the driver pays per miles consumed on the highways? [Probe: charges varying by level of demand, so that rates for miles traveled on some highways would be higher than others, and rates would be higher during the peak hour than in "shoulder" hours --in return, the roads would be less congested]
- What are your questions?
- What concerns you about this new system?
- How do you think this new system would affect your business? More specifically, how would it impact the cost of doing business for you? [PROBE: Consider the effect on the different components of your business: the supply of goods you receive, your shipment of goods, your employees, your customers…]
- Do you think there are any benefits to your business with this new toll-based system versus a tax based system?
- Under what conditions would you accept a toll based system in place of a tax based system?
- What impact would this new system have on the broader business climate?

[Listen for/probe:

- [Explore understanding that pricing can provide free flowing traffic] From what you read, how would this new road pricing system provide free-flowing traffic during peak travel times? Is this believable? How important is this to your business? [listen for/probe: what do you think if the toll rates had to be adjusted periodically to maintain free-flowing traffic]

- [ETC and privacy concerns] What do you think about the electronic tolling technology mentioned in the write-up? [if necessary, probe:] Do you have any issues or concerns? Are your businesses using the E-ZPass or Smart Tag? How does the system work for you?

- [pricing as an environmental strategy] From an environmental point of view, are there any benefits to this new road pricing system? How important are they?

3. <u>Taxes vs. Tolling Part II – Additional Components (10 minutes)</u>
The write-up described some additional features or services that would implemented along with the tolling. These included express bus service, more park-and-ride lots, vanpools and carpools, and flextime. [REFER PARTICIPANTS TO SCENARIO:]

Discuss: From the point of view of your business, what do you think of these additional features of the system? Explore appeal of various components; explore issues or concerns.
- Would you be able to offer your employees flextime, or the option to telecommute?
- Would you be able to help your employees organize vanpools or carpools?

IV. Conclusion (5 minutes)
Final comments, Summary, Thank You's, Good bye's.

MODERATOR'S GUIDE: SHIPPER GROUP

Fairfax, VA and Philadelphia 7/07
(90 minutes total)

I. Introduction (5 minutes)
Focus group ground rules & format
Respondents' self-introductions: name, business
General topic of the discussion – transportation in this area

[NOTE: The following questions/subject areas may not be presented verbatim or in order. All may not be presented in every group and additional issues may be included. Their purpose is to stimulate thought. The discussion will flow naturally, with interventions from the moderator to keep the conversation on track and to probe specific subject areas.]

> The main objective is to assess receptivity to changing from the current tax based transportation system to congestion based/user paid system.

II. Warm Up: Important Issues/Transportation Issues (15 minutes)

Can you describe your business' use of the roads [PROBE: how often are you on the road, what types of roads, how far do you travel, at what times of day]

What transportation issues impact your business? *[Listen for/probe for traffic congestion. Follow up with:]*

What do they think can be done about traffic congestion? Is it solvable problem?

[Listen for/explore – but do not mention]: Gas prices i.e. how have high gas prices affected your business? How have you responded to high gas prices? Pollution/environment]

III. Main Discussion *[building on conversation above]* (65 minutes total)

1. <u>Understanding of Transportation costs and how they are paid for</u> (Max 15 minutes)
Continuing with our discussion on transportation, what's your sense of how current transportation costs are paid for? [**DO NOT PROMPT. Be aware of concern re new VA taxes**]

[IF GAS TAX IS MENTIONED:] Can anyone tell me how much you pay in gas taxes? (i.e. cents/mile)

What do you think of our current system for funding transportation? [PROBE: Do you see any problems with the current funding system?]

2. <u>Taxes Vs Tolling Part I –(40 minutes)</u>

[Explain that these taxes have been insufficient to cover transportation costs, so another method is being considered.] [Present handout with general scenario – participants read]

- Discuss perceptions of the two mechanisms for financing transportation. <u>From the perspective of your business</u>, what are your initial impressions of this new model for funding transportation, where the driver pays per miles consumed on the highways? [Probe: charges varying by level of demand, so that rates for miles traveled on some highways would be higher than others, and rates would be higher during the peak hour than in "shoulder" hours --in return, the roads would be less congested]
- What are your questions?
- What concerns you about this new system?
- How do you think this new system would affect your business? [PROBE: how would it impact the cost of doing business for you?]
- Do you think there are any benefits to your business with this new toll-based system versus a tax based system?
- Under what conditions would you accept a toll based system in place of a tax based system? [PROBE: More truck stops? Dedicated lane on highways for trucks?]
- What impact would this new system have on the broader business climate?

[Listen for/probe:

- [Explore understanding that pricing can provide free flowing traffic] From what you read, how would this new road pricing system provide free-flowing traffic during peak travel times? Is this believable? How important is this to your business? [listen for/probe: what do you think if the toll rates had to be adjusted periodically to maintain free-flowing traffic]

- [ETC and privacy concerns] What do you think about the electronic tolling technology mentioned in the write-up? [if necessary, probe:] Do you have any issues or concerns? Are your businesses using the E-ZPass or Smart Tag? How does the system work for you?

- [pricing as an environmental strategy] From an environmental point of view, are there any benefits to this new road pricing system? How important are they?

IV. Conclusion (5 minutes)
Final comments, Summary, Thank You's, Good bye's.

Appendix B: Scenarios

A New Model for Funding Transportation[6]

Currently transportation is funded through a tax-based system; however, the taxes you pay to fund transportation are insufficient for maintaining the existing roads. In addition, traffic congestion has grown to such an extent that building new roads by itself will not solve the problem of congestion. The U.S. Department of Transportation is considering an alternative way for collecting transportation revenues and for easing traffic congestion during peak travel times. Under this new system, the public would no longer pay the portion of their taxes that funds transportation. Instead, funds for transportation would be collected by charging road users a toll to drive on all freeways (this system would eventually be applied to all major metropolitan areas). The price of the toll would:
- Vary based on location and the time of day (in Northern Virginia, the toll would range from about 25 cents per mile during peak hours down to 0 cents during off-peak hours)
- Be set high enough to allow traffic to flow at the speed limit during peak travel times
- Be adjusted periodically based on changes in traffic patterns so that free-flowing traffic speeds could be maintained

These tolls would be collected using electronic toll collection technology, similar to the system used on the Dulles Toll Road (Route 267). There are no toll booths with this system; all tolls are collected electronically at highway speeds. Road users who have an electronic device called a transponder (for example, the E-ZPass or Smart Tag) would have the toll automatically deducted from a transportation account that they have established. Drivers who do not have a transponder would be "video-tolled." That is, cameras would take pictures of their license plate, and the vehicle owner would be billed by mail for the toll.

As part of this new system, public transportation would be expanded, and express bus service would be provided on freeways across Northern Virginia. New park-and-ride lots would be built to enable drivers to carpool and vanpool more easily. Certified vanpools would not pay a toll, and carpoolers would be able to split the toll among themselves.
In addition, employers would be encouraged to provide more flexible work hours (flextime) and to allow their employees to telecommute (work from home).

In summary, this new user-based road pricing system would replace the current tax-based system for funding transportation. Similar to the way in which you pay for electricity (based on how much and when you use it), drivers would be charged for their use of the roads. In addition to providing transportation revenues, the tolls would be used to encourage some drivers to shift to transit, carpools, or vanpools, to telecommute, or to travel during off-peak hours, including flextime schedules. This would decrease congestion, which would result in fuel consumption savings (i.e. no more sitting in stop-and-start traffic).

---

[6] This scenario was used for the general public groups in Northern Virginia.

A New Model for Funding Roads[7]

The current transportation system does not provide sufficient capacity or quality of service for all who want to use it. Traffic congestion has become a serious problem and will only get worse; building new roads by itself will not solve the problem of congestion. The U.S. Department of Transportation is exploring an alternative way to collect transportation funds and to ease traffic congestion during peak travel times. Under this new approach, taxes would no longer fund roads. Instead, the funds for roads would be raised by charging road users a fee (like a toll) to drive on all freeways (this system would eventually be applied to all major metropolitan areas). The price of the user fee would:

- Vary based on location and the time of day (in Northern Virginia, the toll would range from 0 cents during off-peak hours to about 25 cents per mile during peak hours)
- Be set high enough to allow traffic to flow at the speed limit during peak travel times
- Be adjusted periodically based on changes in traffic patterns so that free-flowing traffic speeds could be maintained

These tolls would be collected using electronic toll collection technology, similar to the system used on the Dulles Toll Road (Route 267). There are no toll booths with this system; all tolls are collected electronically at highway speeds. Road users who have an electronic device called a transponder (for example, the E-ZPass or Smart Tag) would have the toll automatically deducted from a transportation account that they have established. Drivers who do not have a transponder would be "video-tolled." That is, cameras would take pictures of their license plate, and the vehicle owner would be billed by mail for the toll.

In summary, this new user-based road pricing approach would replace the current tax-based approach for funding roads. Similar to the way in which you pay for electricity (based on how much and when you use it), drivers would be charged for their use of the roads. In addition to providing transportation funds, the road user fee would be used to encourage some drivers to shift to transit, carpools, or vanpools, to telecommute, or to travel during off-peak hours, thus decreasing congestion during peak travel times.

---

[7] This scenario was presented to the business and shipper groups in Northern Virginia. It was revised slightly from the original scenario to focus on roads.

US Dept of Transportation Proposes New System for Funding Highways, Local Roads

Would toll all highways at variable rates to reduce congestion.

WASHINGTON, D.C. - The United States Department of Transportation today unveiled a plan to revamp the system that funds all roads and highways in the country.

Interstates and other major highways in metropolitan areas would be tolled at rates varying by time of day to reduce congestion while current taxes that pay for transportation would be eliminated.

Drivers would be encouraged to equip their cars with transponders similar to the EZ Pass used on many highways now. There would be no toll booths because the scanners can work at full highway speeds. Drivers without a transponder would be photographed and receive a bill in the mail with an additional processing fee.

Department of Transportation spokesman Blair Jones states the current system of funding "does not provide sufficient capacity or quality of service for all who want to use our roads and highways." Traffic congestion has "grown to such an extent that building new roads by itself will not solve the problem."

The plan outlined sample tolls for Philadelphia, where tolls on major highways (see map) would be 25 cents per mile during morning and afternoon rush hours, 10 cents a mile during the hour on either side of the rush, and free the rest of the day.

These prices have been set to maintain funding at current levels and to ensure traffic flows at posted speed limits all day. Jones noted that cars no longer stuck in stop-and-start traffic will experience better fuel economy and lower maintenance costs.

Improved Public Transit

The plan also calls for improved public transit options for those wishing to reduce their driving costs. Express bus service, including for those commuting between suburbs, would be expanded, additional park and ride lots will be constructed, carpool ride matching programs would be improved, and certified vanpools will be exempt from tolls. Businesses will also be encouraged to implement or expand telework and flextime programs.

Concerns about Privacy, Costs

Marion Smith of the Citizens Council for the Freedom of Travel, a group opposed to road pricing and favors widening existing highways, claims administrative costs of this new system would balloon out of control, particularly to track down drivers without transponders. Smith added, "Don't even get me started on privacy."

Jones acknowledged the concerns, admitting the high operating costs and that there would be winners and losers, but "Adding just one highway lane in an urban area costs as much as $50 million a mile. That can buy 1,000 miles of 4-lane highway scanners with plenty left over for postage -- and actually get traffic on the whole system to flow freely."

Fast Facts: Road Pricing

- Tolls would range from 25 cents a mile to free, based on time of day.
- The rates would be revised 4 times a year, based on traffic levels.
- Gas taxes and other taxes that fund roads would be eliminated.
- More express buses and park-and-ride lots will be added.
- Carpool, vanpool, telework, and flextime programs expanded.
- Potential to achieve better fuel economy and lower maintenance costs.

Appendix C: Worksheets on Transportation Costs and Toll Costs

Respondent Worksheets

In order to give respondents a sense of how the proposed tolling system would affect them personally, two sets of worksheets were administered during the focus groups: one for estimating current transportation costs and one for estimating potential toll costs under the proposed system. The transportation costs worksheet enabled respondents to calculate how much they currently pay in taxes to fund transportation, and hence what their tax savings would be under the proposed system. The toll costs worksheet enabled respondents to calculate how much they would pay in tolls if congestion pricing was implemented.

1. **Transportation Costs Worksheet**

In developing the transportation costs worksheet, a comprehensive review of transportation funding and expenditures was conducted for both focus group areas. While the worksheets reflected the various ways people are taxed (gasoline, home, vehicle, licenses, etc), these groupings only came about after the collection of budgetary and tax information from four levels of government: federal, state, county, and local. State, county, and local tax rates and transportation appropriations were obtained from state departments of transportation, as well as state, county, and local budgets. The budgetary and tax information used to develop the worksheets are detailed below.

Federal Taxes
The federal fuel excise tax is the major source of transportation funding at the federal level. Currently, virtually all funds go into either the Highway Trust Fund or the Transit Trust fund, so the tax of 18.4 cents per mile was included in the calculations.

State Taxes
The fuel tax is also a major source of state transportation funding. The entirety of the tax is maintained within the state department of transportation and was included in the calculations. Northern Virginia also charges a 2.5% sales tax on gasoline to fund regional transit; this amount was also included in the estimates with an assumed price of $2.75 per gallon before taxes.

Vehicle title and registration fees as well as driver's license fees are the other major source of state transportation funds.[8] These fees were included in the worksheet calculations. For the worksheet, to estimate approximate annual costs, yearly fees, such as for license plates or drivers licenses were amortized over the life of license, if valid for greater than one year. Titling fees were divided evenly over the life of the vehicle.

Virginia
Vehicle sales taxes (2%) are retained within the state department of transportation. For the worksheet, the tax was divided evenly over the life of the vehicle.

---

[8] The final major component is income from fines.

Pennsylvania
The above taxes do not provide enough funding to maintain the department of transportation alone. As a result, the department of transportation receives a transfer from the state's General Fund. The major contributions to the General Fund are: state income taxes and sales taxes (including sales tax on vehicles). Only about 1% of the General Fund is directed to the department of transportation and so only 1% of estimated income and sales taxes are added into the transportation costs on the worksheet.

County Taxes
Virginia
Taxes rates vary across counties but all taxes fit into the same categories. All the Northern Virginia counties charge sales tax, real property, and personal property taxes. These make up the largest portion of contributions to the county's General Fund, which then funds county streets and roads. The percentage of the General Fund allocated to transportation is usually around 1% (with the exception of the independent city of Alexandria, where the rate is 8%). The worksheets are designed to apply to people from throughout the area so the estimated taxes are calculated using an approximate average rate for the region.

Pennsylvania
With the exception of the independent city of Philadelphia, Pennsylvania counties collect little in the way of taxes and spend even less on transportation. The only expenditure of note is a yearly subsidy to SEPTA, the transit agency. Property taxes are the only significant revenue source for the counties, of which portions from 2-20% go to SEPTA and so an "average" value is calculated for use in the worksheets.

Cities
Virginia
Cities in Virginia collect real property and personal property taxes on vehicles that go into the General Fund. They also sell annual parking permits to residents. The percentage of the General Fund that goes to roads and streets ranges from 3-12%. One city does not collect personal property tax and so estimates for residents of Herndon will be slightly high on this line of the worksheet.

Pennsylvania
Pennsylvania cities (except Philadelphia) can collect income and real property taxes for the General Fund, of which between 1-13% goes toward transportation. For use in the worksheets, an "average" rate for income and real property were used, though with the caveat that the rates of each vary greatly in the metropolitan area (i.e. some cities do not collect income tax) but that the total collection between property and income taxes is fairly constant from city to city. Philadelphia collects a local sales tax and has income and property taxes at rates higher than the surrounding areas. City residents will likely have their total bill underestimated.

Developing the Worksheet

Based on the budgetary and tax information collected for Pennsylvania and Virginia, separate worksheets were developed for each site. The worksheets had to be simple and clear, with a minimum number of calculations, so that respondents could complete them independently and with ease. To reduce the number of calculations, federal, state, county, and city taxes were grouped together, as appropriate, by "taxable object" (i.e., gasoline, home, vehicle, etc.). The following two sections provide additional details on the development of the worksheet for the two focus group sites.

Northern Virginia

Six tax groupings were identified in Northern Virginia, including fuel taxes, taxes on your home, taxes on your car, sales tax, taxes on new vehicles and yearly vehicle fees. For each of these groupings, tables were developed that provided an estimate of the tax paid, based on different valuations or respondent behaviors. For example, regarding fuel taxes, the table included eight separate ranges for number of miles driven in a year, (e.g., 6,000 – 7,999 miles per year, 8,000- 9,999 miles per year, 10,000 – 11,999 miles per year, etc). For each of these ranges, the approximate amount of tax paid (based on average miles per gallon for a medium sized car) was indicated on the table.[9] Respondents simply had to look up how much tax they paid based on their number of miles driven per year, and they wrote the amount on the worksheet. Respondents looked up the taxes paid for each of the tax groupings, and they summed those figures (using a calculator) to arrive at an estimate for the annual amount of taxes they pay to fund transportation.

The following table highlights the behavior or value associated with each tax grouping, as well as the underlying taxes that were included in the table.

Northern Virginia		
Tax Groupings	**Behavior or Valuation**	**Underlying Taxes**
Fuel Taxes	Miles driven per year	• Federal fuel tax • State fuel tax • Northern Virginia fuel tax surcharge
Taxes on your home	Home value	• County real property tax • Local real property tax
Taxes on your car	Car value (current)	• County personal property tax • Local personal property tax
Sales tax	Income	• County Sales tax
Taxes on new vehicles	Car value (when purchased)	• Vehicle sales tax • Vehicle title
Yearly vehicle fees	NA (fixed cost)	• Vehicle registration • Driver's license fee

---

[9] A similar table for sport utility vehicles was included as well.

Philadelphia

The general tax groupings identified for Philadelphia were similar to Northern Virginia, except that in Philadelphia there are no personal property taxes (i.e. taxes on your car), and in Philadelphia a portion of the income tax funds transportation. The following table highlights the key elements of the worksheet designed for Philadelphia.

Philadelphia		
Tax Groupings	**Behavior or Valuation**	**Underlying Taxes**
Fuel Taxes	Miles Driven per year	• Federal fuel tax • State fuel tax
Taxes on your home	Home value	• County real property tax • Local real property tax
Earned income taxes	Income	• County personal property tax • Local personal property tax
Sales tax	Income	• County Sales tax
Taxes on new vehicles	Car value (when purchased)	• Vehicle sales tax • Vehicle title
Yearly vehicle fees	NA (fixed cost)	• Vehicle registration • Driver's license fee

2. **Toll Costs Worksheet**

For the toll costs worksheet, respondents were asked to think about the number of miles they drive on the highway during high peak, or rush hour (defined as 7 a.m. to 8 a.m. and 5 p.m. to 6 p.m. on weekdays), as well as the number of miles they drive on the highway during low peak (defined as the hour on either side of high peak on weekdays and Saturdays from 10 a.m. to 2 p.m.). Two tables were included on the worksheet, one showing toll costs during high peak and one showing toll costs during low peak. In order to simplify the task for the respondent, the two tables showed toll costs associated with different ranges of miles driven (e.g., 0-10 miles, 11-20 miles, 21-30 miles etc); for each range the mean toll cost was calculated (and rounded to the nearest dollar). In calculating the toll costs for the tables, a toll rate of 25 cents per mile was used for high peak and 15 cents per mile was used for low peak. Based on the number of highway miles driven during high peak, respondents could look up the associated toll cost in the table (and similarly for highway miles driven during low peak). For example, if a respondent drives 25 highway miles during peak, she would find that the toll associated with 21-30 miles (during peak) is $6.

Potential Toll Costs Worksheet[10]

For our example, tolls will be established at three rates. 1) High peak or "rush hour" 2) Low peak, the times on either side of rush hour, and 3) Saturday low peak. This sheet will help you calculate the amount you would pay in tolls if you do not change your travel behavior. In each section, if you do not make the specified trip, enter a zero in the box.

A. Freeway Commute (High Peak, Monday-Friday)
In Table 1, look up how many miles you commute to and from work on freeways and tollways **on a weekly basis between 7am and 8am as well as between 5pm and 6pm**. Enter the cost in the next column in Box A. If your work location or schedule varies, use an average.

BOX A
$

+

B. Freeway Non-Commute (High Peak, Monday-Friday)
In Table 2, look up how many miles you travel (on average) on freeways and tollways when making non-work related trips (i.e. shopping, medical appointments) **each week between 7am and 8am as well as between 5pm and 6pm**. Enter the cost in Box B.

BOX B
$

+

C. Freeway Commute (**Low** Peak, Monday-Friday)
In Table 1, look up how many miles you commute to and from work on freeways and tollways **on a weekly basis between 6am and 7am, between 8am and 9am, as well as between 3pm and 5pm and between 6pm and 7pm.** Enter the cost in Box C. If your work location or schedule varies, use an average.

BOX C
$

+

D. Freeway Non-Commute (**Low** Peak, Monday-Friday)
In Table 2, look up how many miles you travel (on average) on freeways and tollways when making non-work related trips (i.e. shopping, medical appointments) **each week between 6am and 7am, between 8am and 9am, as well as between 3pm and 5pm and between 6pm and 7pm.** Enter the cost in Box D.

BOX D
$

SUBTOTAL
Add together the numbers in Boxes A, B, C, and D and enter the value in Box 1.

BOX 1
$

---
[10] This same toll costs worksheet was used in Philadelphia, with appropriate changes made to the heading.

SUBTOTAL
Re-enter the value from Box 1 here.

BOX 1
$

\+

E. Saturday Freeway Trips (**Low** Peak)
In Table 2, look up how many miles you travel, on average, on freeways and tollways on **Saturday from 10am to 2pm.** Enter the cost in Box E.

BOX E
$

Weekly Total
Add together the values from Box 1 and Box E and enter the value in Box 2.

BOX 2
$

x

48

ANNUAL TOTAL
Multiply the value from Box 2 by 48 (the approximate number of weeks a year you would be making these trips) and enter the value in Box 3.

BOX 3
$

Virginia Toll Cost Tables

Table 1 High Peak ("Rush Hour")	
Miles	Cost
0 – 10	$1
11 – 20	$4
21 – 30	$6
31 – 40	$9
41 – 50	$11
51 – 60	$14
61 – 70	$16
71 – 80	$19
81 – 90	$21
91 – 100	$24
101 – 120	$28
121 – 140	$33
141 – 160	$38
161 – 180	$43
181 – 200	$48
201 – 220	$53
221 – 240	$58
241 – 260	$63
261 – 280	$68
281 – 300	$73
301 – 320	$78
321 – 340	$83
341 – 360	$88
361 – 380	$93
381 – 400	$98
401 – 420	$103
421 – 440	$108
441 – 460	$113
461 – 480	$118
481 – 500	$123
501+	$128

Table 2 Low Peak (M-F, Sat)	
Miles	Cost
0 – 10	$1
11 – 20	$2
21 – 30	$4
31 – 40	$5
41 – 50	$7
51 – 60	$8
61 – 70	$10
71 – 80	$11
81 – 90	$13
91 – 100	$14
101 – 120	$17
121 – 140	$20
141 – 160	$23
161 – 180	$26
181 – 200	$29
201 – 220	$32
221 – 240	$35
241 – 260	$38
261 – 280	$41
281 – 300	$44
301 – 320	$47
321 – 340	$50
341 – 360	$53
361 – 380	$56
381 – 400	$59
401 – 420	$62
421 – 440	$65
441 – 460	$68
461 – 480	$71
481 – 500	$74
501+	$77

Current Transportation Cost Worksheet: Virginia

Streets, Highways, and Transit are funded by a variety of sources in Virginia. In order to understand how much you are paying today, complete this worksheet with the help of the attached tables.

A. Fuel Taxes
Northern Virginia charges a tax for every gallon of fuel purchased. Look up the number of miles you drive per year on Table 1 if you drive a car and Table 2 if you drive an SUV. Enter the corresponding number in Box A.

BOX A
$

+

B. Taxes on Your Home
Homeowners pay real property taxes to their county and city of residence. Look up the current value of your home in Table 3 and enter the value next to it in Box B.

BOX B
$

+

C. Taxes on Your Car
Each year vehicle owners pay personal property taxes on their vehicles to their county and city of residence. Look up the current value of your vehicle in Table 4 and enter the value next to it in Box C.

BOX C
$

+

D. Sales Taxes
A portion of state and local sales taxes are spent on transportation. Look up your current salary in Table 5 and enter the estimate in Box D.

BOX D
$

SUBTOTAL
Add together the numbers in Boxes A, B, C, and D and enter the value in Box 1.

BOX 1
$

Re-enter the information from Box 1 here.

BOX 1
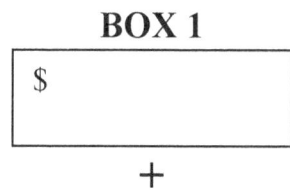

+

Taxes on New Vehicles
Virginia residents pay sales tax and titling fees on vehicles (used or new) when purchased. Look up the value of your vehicle (when purchased) in Table 6 and enter the estimate of sales taxes and titling fees in Box E.

In Box F, enter how many years you usually go between buying a new car.

Divide the number in Box E by the number in Box F and put the result in Box G.

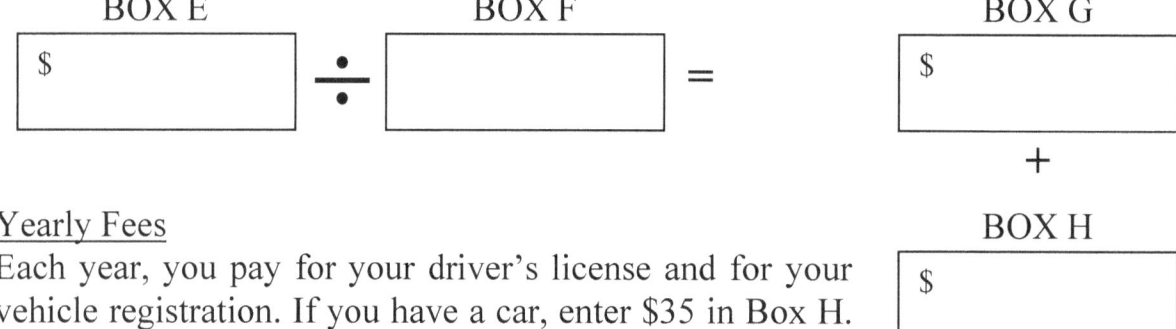

+

Yearly Fees
Each year, you pay for your driver's license and for your vehicle registration. If you have a car, enter $35 in Box H. If you have an SUV. enter $40 in Box H.

BOX H
$

TOTAL
Add together the numbers in Box 1, Box G, and Box H in Box 2.

BOX 2 (TOTAL)
$

Virginia Current Transportation Cost Tables

Please use the tables on this page to estimate the amount of your taxes that fund transportation.

TABLE 1 Car Fuel Tax	
Miles	Tax
6,000 – 7,999	$150
8,000 – 9,999	$190
10,000 – 11,999	$240
12,000 – 13,999	$280
14,000 – 15,999	$320
16,000 – 17,999	$370
18,000 – 19,999	$410
20,000+	$430

Table 2 SUV Fuel Tax	
Miles	Tax
6,000 – 7,999	$190
8,000 – 9,999	$240
10,000 – 11,999	$300
12,000 – 13,999	$350
14,000 – 15,999	$410
16,000 – 17,999	$460
18,000 – 19,999	$510
20,000+	$540

Table 3 Real Property Tax	
Home Value	Tax
$75,000 - $149,000	$50
$150,000 - $249,000	$90
$250,000 - $349,999	$140
$350,000 - $449,000	$180
$450,000 - $599,999	$240
$600,000 - $799,000	$320
$800,000 - $999,999	$410
$1,000,000 - $1,199,999	$500
$1,200,000 - $1,399,999	$590
$1,400,000 - $1,599,999	$680
$1,600,000+	$770

Table 4 Personal Property Tax	
Car Value (Current)	Tax
$1,000 - $4,999	$0
$5,000 - $14,999	$10
$15,000 - $24,999	$20
$25,000 - $34,999	$30
$35,000 - $44,999	$40
$45,000 - $54,999	$50
$55,000 - $64,999	$60
$65,000 - $74,999	$70
$75,000 - $84,999	$80
$85,000 - $94,999	$90
$95,000 - $104,999	$100

Table 5 Sales Taxes	
Income	Tax
$20,000 - $49,999	$30
$50,000 - $79,999	$60
$80,000 – $109,999	$90
$110,000 - $139,999	$120
$140,000 - $169,999	$150
$170,000 - $199,999	$180
$200,000+	$210

Table 6 Vehicle Sales Taxes	
Car Value (New)	Tax
$1,000 - $4,999	$100
$5,000 - $9,999	$250
$10,000 - $14,999	$400
$15,000 - $19,999	$550
$20,000 - $24,999	$700
$25,000 - $34,999	$910
$35,000 - $44,999	$1,210
$45,000 - $54,999	$1,510
$55,000 - $64,999	$1,810
$65,000 - $79,999	$2,200
$80,000 - $99,999	$2,710

Current Transportation Cost Worksheet: Pennsylvania

Streets, Highways, and Transit are funded by a variety of sources in Pennsylvania. In order to understand how much you are paying today, complete this worksheet with the help of the attached tables.

PLEASE NOTE: The taxes you pay vary from municipality to municipality. However, the overall tax payment is usually the same. For example, places with no income tax often have a higher property tax. Regardless of where you live, fill out the worksheet as instructed and the results will be a reasonable estimate of your current total costs.

A. Fuel Taxes
Pennsylvania and the Federal government charge a tax for every gallon of fuel purchased. Look up the number of miles you drive per year on Table 1 if you drive a car and Table 2 if you drive an SUV. Enter the corresponding number in Box A.

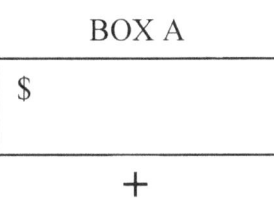

BOX A
$

+

B. Taxes on Your Home
Homeowners pay property taxes to their county and city of residence. Look up the current value of your home in Table 3 and enter the value next to it in Box B.

BOX B
$

+

C. Earned Income Taxes
Each year residents pay earned income taxes to the state and (sometimes) city of residence. Look up your current salary in Table 4 and enter the value next to it in Box C.

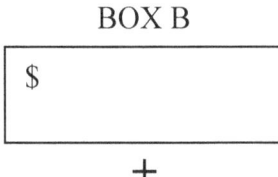

BOX C
$

+

D. Sales Taxes
A portion of state and local sales taxes are spent on transportation. Look up your current salary in Table 5 and enter the estimate in Box D.

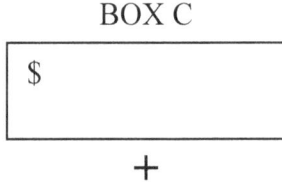

BOX D
$

SUBTOTAL
Add together the numbers in Boxes A, B, C, and D and enter the value in Box 1.

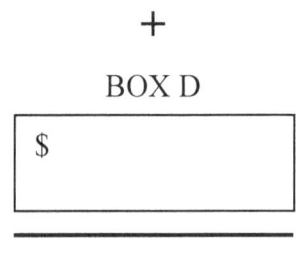

BOX 1
$

Re-enter the information from Box 1 here.

BOX 1
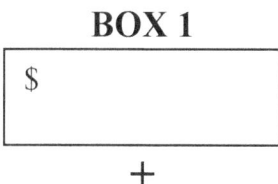

+

Taxes on New Vehicles
Pennsylvania residents pay sales tax and titling fees on vehicles (used or new) when purchased. Look up the value of your vehicle (when purchased) in Table 6 and enter the estimate of sales taxes and titling fees in Box E.

In Box F, enter how many years you usually go between buying a new car.

Divide the number in Box E by the number in Box F and put the result in Box G.

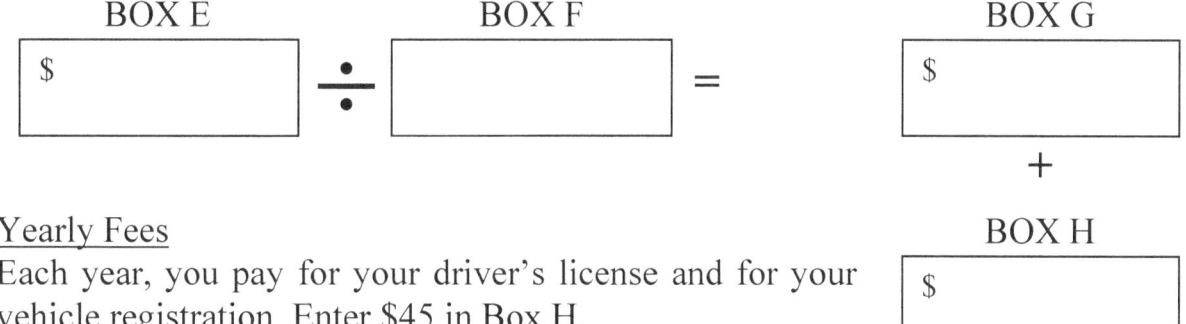

+

Yearly Fees
Each year, you pay for your driver's license and for your vehicle registration. Enter $45 in Box H.

BOX H
$

BOX 2 (TOTAL)
$

TOTAL
Add together the numbers in Box 1, Box G, and Box H in Box 2.

57

Pennsylvania Current Transportation Cost Tables

Please use the tables on this page to estimate the amount of your taxes that fund transportation.

TABLE 1	
Car Fuel Tax	
Miles	Tax
6,000 – 7,999	$170
8,000 – 9,999	$220
10,000 – 11,999	$270
12,000 – 13,999	$320
14,000 – 15,999	$370
16,000 – 17,999	$420
18,000 – 19,999	$470
20,000+	$500

Table 2	
SUV Fuel Tax	
Miles	Tax
6,000 – 7,999	$220
8,000 – 9,999	$280
10,000 – 11,999	$340
12,000 – 13,999	$400
14,000 – 15,999	$450
16,000 – 17,999	$530
18,000 – 19,999	$590
20,000+	$620

Table 3	
Real Property Tax	
Home Value	Tax
$75,000 - $149,000	$110
$150,000 - $249,000	$180
$250,000 - $349,999	$270
$350,000 - $449,000	$360
$450,000 - $599,999	$470
$600,000 - $799,000	$630
$800,000 - $999,999	$810
$1,000,000 - $1,199,999	$990
$1,200,000 - $1,399,999	$1,170
$1,400,000 - $1,599,999	$1,350
$1,600,000+	$1,530

Table 4	
Earned Income Tax	
Income	Tax
$20,000 - $49,999	$40
$50,000 - $79,999	$70
$80,000 – $109,999	$100
$110,000 - $139,999	$140
$140,000 - $169,999	$170
$170,000 - $199,999	$200
$200,000+	$230

Table 5	
Sales Taxes	
Income	Tax
$20,000 - $49,999	$10
$50,000 - $79,999	$10
$80,000 – $109,999	$10
$110,000 - $139,999	$20
$140,000 - $169,999	$20
$170,000 - $199,999	$30
$200,000+	$30

Table 6	
Vehicle Sales Tax & Title	
Car Value (New)	Tax
$1,000 - $14,999	$30
$15,000 - $24,999	$40
$25,000 - $34,999	$50
$35,000 - $44,999	$60
$45,000 - $54,999	$60
$55,000 - $64,999	$70
$65,000 - $79,999	$80
$80,000 - $99,999	$100

Appendix D: Questions Raised by Participants

The following table summarizes the questions raised by respondents in each of the focus groups. Many of these questions or topics are addressed in the findings of this report; they are included here to provide a flavor for respondents' interest and concerns regarding the congestion pricing initiative.

Northern Virginia	Philadelphia
General Public	**General Public**
<ul><li>Has it been implemented somewhere else?</li><li>Who is going to certify vanpools? Is there a fee?</li><li>What about other roads – would they still be maintained?</li><li>How to handle people who don't live here?</li><li>What is the incentive to get an E-ZPass?</li><li>How do you know the roads will be less congested?</li><li>Where's the budget? Where are the current revenues coming from and how will they get replaced? What are the real mechanics behind this?</li><li>How can they do it? How can they take away all this funding and only charge for driving on freeways?</li><li>What will they do with the funding [from tolls]?</li><li>Where has this worked?</li><li>How to manage non-toll roads?</li><li>Will reverse commute be charged?</li></ul>	<ul><li>What is rush hour?</li><li>What will they consider a metropolitan area?</li><li>What about truck drivers (gas already killing them)?</li><li>Will it just shift congestion to a new hour?</li><li>How do they get tax dollars from folks who never go on toll roads?</li><li>Can we change it back [i.e., if system doesn't work]?</li><li>Is there a discount with E-ZPass for seniors?</li><li>If taxes are eliminated does that offset the cost of riding on the highway?</li><li>Do you go through something? Will you know when you are being charged?</li><li>Wouldn't everyone change their hours to avoid the toll?</li><li>How would this affect businesses?</li><li>How can everyone get a transponder?</li><li>Why not institute some of the other things? Why not carpool lanes?</li><li>What next? Tax local roads?</li></ul>

Northern Virginia	Philadelphia
Business Group	**Business Group**
<ul><li>What about people driving through?</li><li>Which taxes are disappearing?</li><li>How far will tax reach? Just Virginia?</li><li>Cost to collect/print bill – who's paying?</li><li>What would be the timeline to get the technology?</li><li>What is the purpose?</li><li>How much money to enforce?</li><li>How can all the roads be less congested?</li><li>Does each jurisdiction track billing? How does it work?</li><li>How to decide what to do with the money they get?</li><li>Will money be spent appropriately?</li><li>Would my transponder work in all metropolitan areas?</li></ul>	<ul><li>What is "rush hour"?</li><li>What about privacy?</li></ul>
Shipper Group	
<ul><li>Is this to generate more money? Or a burden on commercial vehicles?</li><li>Cost in running the program – what will be left for the roads?</li><li>Benefit us or out-of-state?</li><li>Is this supposed to be revenue neutral?</li><li>What do they want with this?</li><li>On what roads?</li><li>How do they define off-peak?</li><li>What's the cost going to be?</li></ul>	<ul><li>What about people who don't use credit?</li><li>Will gas prices go down at every pump in the United States?</li><li>What will it cost to run this?</li><li>What's to discourage me from taking grandma's car?</li><li>How will congestion be reduced?</li><li>What agency will handle this?</li></ul>

www.ingramcontent.com/pod-product-compliance
Lightning Source LLC
Chambersburg PA
CBHW081852170526
45167CB00007B/2977